"A must-read for anyone who has wondered how we can maintain our humanity amid the superpowerful prediction machines we've created."

—**ANGELA DUCKWORTH,** author, *New York Times*–bestselling *Grit*

"A compelling read about how AI is shaping us—and how we should shape it. Tomas Chamorro-Premuzic examines how technology can augment our intelligence and reminds us to invest in the human skills that robots can't replace."

—**ADAM GRANT,** author, number one *New York Times*–bestselling *Think Again*; host, TED podcast *Re:Thinking*

"At last, a book on AI that focuses on humans rather than machines. A powerful case for reclaiming some of our most valuable neglected virtues."

—**DORIE CLARK,** author, *Wall Street Journal*–bestselling *The Long Game*; executive education faculty, Duke University Fuqua School of Business

"Techno-zealots and doomsayers dominate the debate about artificial intelligence, which is why this unique book is such a breath of fresh air. *I, Human* is a strikingly clear-eyed account of the fraught but fertile relationship we already have with AI—and an inspiring argument for how, in the future, it can help us maintain and enhance rather than degrade what makes us essentially human."

—**OLIVER BURKEMAN,** author, *New York Times*–bestselling *Four Thousand Weeks*

"If you want to understand how we can best thrive in a world that is rapidly changing because of AI, and feel hopeful and confident about the role you can play, you'll find this book to be both brilliant and essential. Full of insights and practical tips, *I, Human* will prepare you for the future by focusing your attention on the very traits that make human nature unique."

—**FRANCESCA GINO,** professor, Harvard Business School; author, *Rebel Talent*

"*I, Human* argues compellingly that artificial intelligence is altering human intelligence—fueling narcissism, diluting self-control, reinforcing prejudice—and reveals how human learning can still counteract the malign effects of machine learning. Tomas's easy style and dry humor bely the seriousness with which he tackles this vital issue of our time. Take note before the robots take over how you think."

—**OCTAVIUS BLACK,** founder and CEO, MindGym

"As someone whose own work is all about using technology to make good things happen in the real world, I could not welcome Tomas Chamorro-Premuzic's excellent treatise on AI and how it relates to our collective humanity more. This book is a must-read for anyone worried about the implications of AI, because it focuses not just on the evolution of artificial intelligence but on how we can evolve and upgrade our own intelligence as a result. I guarantee you'll come away inspired, optimistic about humanity's future, and empowered in your own ability to manage and define it."

—**CINDY GALLOP,** founder and CEO, MakeLoveNotPorn

"Dr. Tomas takes our knowledge of AI in an entirely new direction by helping us better understand both the machines and ourselves. A book that will spur you to live better and learn more."

—**JULIA GILLARD,** former Prime Minister of Australia; Chair, Wellcome Trust and the Global Institute for Women's Leadership

"As AI becomes more prevalent, we all need to be more sensitive to how it changes our behavior. Tomas's book raises important issues we all have to consider as AI expands into every part of our daily lives."

—**JOSH BERSIN,** global industry analyst

"In this thought-provoking book, Tomas Chamorro-Premuzic trains his psychologist's eye on what makes us truly human—warts and all. Packed with illuminating insights about our defining strengths and foibles *I, Human* offers a road map for prospering in a world saturated by smart technology."

—**HERMINIA IBARRA,** Charles Handy Professor of Organisational Behaviour, London Business School

"Tomas Chamorro-Premuzic's authoritative look at AI's grip on our lives is as entertaining as it is informative—that is to say, it is both. The book is sprinkled with just enough humor and irreverence to make its technical topic engaging and more than enough evidence to make it compelling. It details how algorithms are hijacking our attention, making us more impatient and arrogant and less capable of focus, deep thinking, and ingenuity—and less happy. Ultimately, *I, Human* seeks to awaken our better angels so that we can reclaim meaningful, fulfilling lives of choice and connection."

—**AMY C. EDMONDSON,** professor, Harvard Business School; author, *The Fearless Organization*

"A hugely insightful and timely guide. Dr. Tomas opens a bold window to the impact of artificial intelligence on human behavior. This is your all-things-necessary if you care about humanizing work in this highly technological age. You most likely know why this is wildly important to you, but struggle with how to do it. This book is an excellent read as we attempt to build a better working world."

—**KATARINA BERG,** CHRO, Spotify

I, Human

I, Human

AI, Automation, and the Quest to Reclaim What Makes Us Unique

TOMAS CHAMORRO-PREMUZIC

Harvard Business Review Press

Boston, Massachusetts

The web addresses referenced in this book were live and correct at the time of the book's publication but may be subject to change.

Library of Congress Cataloging-in-Publication Data

Names: Chamorro-Premuzic, Tomas, author.
Title: I, human : AI, automation, and the quest to reclaim what makes us unique / Tomas Chamorro-Premuzic.
Description: Boston, Massachusetts : Harvard Business Review Press, [2022] | Includes bibliographical references and index. |
Identifiers: LCCN 2022034376 (print) | LCCN 2022034377 (ebook) | ISBN 9781647820558 (hardcover) | ISBN 9781647820565 (epub)
Subjects: LCSH: Technology and civilization. | Technology--Social aspects. | Artificial intelligence. | Human beings--Effect of technological innovations on. | Humanity. | Emotional intelligence.
Classification: LCC CB478 .C4936 2022 (print) | LCC CB478 (ebook) | DDC 303.48/3--dc23/eng/20220830
LC record available at https://lccn.loc.gov/2022034376
LC ebook record available at https://lccn.loc.gov/2022034377

ISBN: 978-1-64782-055-8
eISBN: 978-1-64782-056-5

The paper used in this publication meets the requirements of the American National Standard for Permanence of Paper for Publications and Documents in Libraries and Archives Z39.48-1992.

For Isabelle and Viktor,
to a life rich in analogue adventures.

This book examines the impact of artificial intelligence on human behavior.

Contents

Introduction   1

Chapter 1: Being in the AI Age   7

Chapter 2: Weapons of Mass Distraction   29

Chapter 3: The End of Patience   47

Chapter 4: Taming Bias   61

Chapter 5: Digital Narcissism   83

Chapter 6: The Rise of Predictable Machines   103

Chapter 7: Automating Curiosity   121

Chapter 8: How to Be Human   141

Notes   159
Index   175
Acknowledgments   185
About the Author   187

I, Human

Introduction

A lot of ink and bytes have been devoted to the rise of AI. The business press and experts have heralded AI as the next big thing fueling the new industrial revolution. Techno-evangelists claim that the technology will transform our jobs, help cure diseases, and extinguish all human biases.

I'm also sure you've heard the bleak predictions, some by the world's most eminent techno-enthusiasts, that AI threatens to upend the human species as we know it. For example, the usually optimistic Bill Gates confessed that he is "concerned about super intelligence." Likewise, the late Stephen Hawking noted that "a super-intelligent AI will be extremely good at accomplishing its goals, and if those goals aren't aligned with ours, we're in trouble."[1] Meanwhile, Elon Musk labeled AI "a fundamental risk to the existence of human civilization," though that hasn't stopped him from trying to implant it in our brains.

So, you may be wondering why you need to read yet *another* book on AI. Even short books like this one are a considerable investment of time, energy, and focus, all of which are precious and scarce. I'll explain why.

Despite all the forecasts about AI—from tech-utopians and Luddites alike—one topic has been oddly neglected: how AI is changing our lives, values, and fundamental ways of being. It is time to look at AI from a human perspective, which ought to include an assessment of how the AI age is impacting human behavior. How is AI changing the way we work, as well as other areas of life, such as relationships, well-being, and consumption? What are the key social and cultural differences between the AI era and previous chapters in human civilization? And how is AI redefining the key forms in which we express our humanity?

These questions fascinate me. As a psychologist, I've studied human traits and foibles for decades, trying to understand what makes us tick. For over twenty years, much of my research has focused on understanding human intelligence: how to define it and measure it, and what happens when we decide not to use it, especially when we select leaders. This research has highlighted how our flaws and misconceptions shape our world, generally not for the better: our overreliance on intuition over data, our tendency to mistake confidence for competence, and our propensity to prefer incompetent male leaders over competent female (and competent male) leaders all account for many of the major challenges we face in the world. And, as a scientist-practitioner, I've spent my career trying to find ways to help people and organizations make better, data-driven decisions. This is how I initially stumbled upon AI, namely as a tool that has clear potential for decoding the dynamics of people at work and not just predict, but also enable, better performance for individuals, teams, and organizations. I devote much of my time to designing and deploying AI to select the right employees, managers, and leaders, and to increase diversity and fairness in organizations, so that more people, and particularly those who have historically been unfairly excluded, can thrive at work.[2]

Since nobody has any data on the future, it is hard to know how AI will unfold. At least until now, AI has been mostly just something that happens to data. It usually boils down to partially

self-generated algorithms that possess a relentless capacity for iden-
tifying hidden patterns in a big set of data, via their ability to evolve,
learn, unlearn, autocorrect, and perfect, regardless of whether they
end up reaching (or surpassing) human levels of intelligence.

Yet, its stealth omnipresence is impactful. Whether we're cog-
nizant of it or not, we interact with AI daily: When we ask Siri or
Alexa a question. When we're exposed to a digital ad. When we're
shown news or any content online. Given how much we use our
phones and scroll through social media, we probably spend more
time interacting with AI than we do with our spouses, friends, and
coworkers, all of whom are also influenced by AI when they interact with us. AI is omnipresent, and though it's still evolving and so
much is uncertain, there's no doubt: it's also redefining our lives,
our interactions with the world, and ourselves.

AI has the *potential* to improve our lives. We live in a complex
world, and our archaic brains can no longer rely on intuitive or in-
stinctive decisions to make the right choices, especially if we want
to be functional members of modern society. For example, we can
expect well-designed AI to do a better job than most human recruiters at evaluating the résumé or interview performance of a job
applicant, just as we can expect well-designed AI to outperform
most human drivers; make more accurate, reliable, and faster medical diagnoses than the naked human eye; and outperform humans
at detecting credit card fraud.[3] *Human* bias permeates every aspect
of life, from who gets hired and promoted to who has access to
credit, loans, and college and who gets convicted and incarcerated.

Meritocracy—the idea that our fate should be determined by our
level of skill and effort—is a near-universal aspiration, yet anywhere
in the world it is caused more by privilege and class than other factors.[4] Your birthplace, your parents, and your sociodemographic

classification are all stronger predictors of your future success than your actual potential and performance, especially in the United States. More than any other technological invention, AI has the ability to expose these biases, as well as identify real signals of talent and potential while being fully agnostic to our class, gender, race, and status. Importantly, the key goal is not for AI to replace human expertise, but to enhance it. In any area of decision-making, human expertise will improve with the help of data-driven insights produced by AI.

But the AI age has also unleashed bad behavioral tendencies, which we'll uncover over the course of this book. The deployment of algorithms that co-opt or hijack our attention are contributing to a crisis of distractibility. The AI age is also making us more impatient, ignorant, and delusional, reinforcing our self-serving interpretations of the world. It has also increased our addiction to social media platforms, which have democratized digital narcissism and turned the AI age into an age of self-obsession, entitlement, and ubiquitous attention-seeking. Furthermore, the AI age has also turned us into rather more boring and predictable creatures, diluting the range and richness of experiences that once characterized human life. And, last, AI may be diminishing our intellectual and social curiosity, feeding us rapid and simple answers to everything and discouraging us from actually asking questions.

Perhaps things will get better in the future. After all, AI is still in its infancy, and one would hope that its evolution will also include our own ability to manage and deal with it better so that we can reap the benefits of technological progress. However, so far there are clearly reasons to worry about the behavioral impact and repercussions of the AI age. My goal with this book is to talk about the present rather than the future, focusing on the current realities of humans interacting with AI.

Humans have a strong track record of blaming their technological inventions for their own cultural demise and degeneration. Since the dawn of television, critics have repeatedly blamed TV for being "the opium of the people," inhibiting human imagination and intellectual development, and fueling violence and aggression. When the first newspapers began to circulate in the sixteenth century, skeptics feared they would forever kill social gatherings: Why bother meeting, if there is no more news or gossip to exchange? Long before that, Socrates, like many of his fellow philosophers, refrained from writing altogether on the basis that it would atrophy his memory.[5]

This solid history of slamming novel media tools may seem sufficient to dismiss alarmist criticisms of current technological innovations, though a default underreaction is not necessarily the best alternative to an overreaction. Since one of the key advantages of the AI age is the chance to gather and analyze a great deal of detailed data on human behavior, why not harness the opportunity to assess the impact of the AI age on human behavior in an evidence-based way?

That's what I aim to do: focus less on the potential future consequences and more on what has happened thus far—worrying less about future threats and more about present realities.

With that in mind, my aim for this book is to ask big questions: What does it mean to be human in the AI age, and in what new and perhaps better ways could we express our humanity in this current chapter of our evolutionary history? Amid much discussion on how the rise of AI is taking control of the world and our lives, can we reclaim our humanity to display our most virtuous side and avoid being alienated or dehumanized, let alone automated, by technology?

Given that AI is still evolving, these questions may yet be unanswerable, but this shouldn't stop us from trying to answer them. Even while the AI age unfolds, we can reflect on what we are observing at this very point in time of the human-AI interface.

You never see yourself age in the mirror, but one day you find an old picture of yourself and realize that you have changed. Along the same lines, if we obsess too much about the future, we will risk neglecting the present. Rather than adding to the overly saturated world of technological predictions, let us instead look at our present to understand where we are and how it is that we got here in the first place. Incidentally, this is also the best way to understand, or at least reflect on, where we may be heading. If we don't like what we see, we will at least have an incentive to change it.

Chapter 1

Being in the AI Age

What AI is and is not

We design tech and tech, in turn, designs us.

—Pamela Pavliscak

In the approximately three-hundred-thousand-year existence of what are generally considered anatomically modern humans, we haven't fundamentally changed all that much. There's no significant biological difference between the present AI pioneers and their ancestral relatives who invented agriculture or any of the major innovation breakthroughs in history. The most advanced and up-to-date version of our species, which includes Angela Merkel, Beyoncé, Jeff Bezos, and me (which I hope is not interpreted as a narcissistic statement), still shares about 99 percent of its DNA with chimpanzees.

Our wants and needs haven't changed that much either. But how those wants and needs are manifested can change with the times. In our evolutionary journey, we managed to transition from producers and users of rudimentary hunter-gathering tools to creators of space rockets, Bitcoin, and RNA vaccines. In the process, we created a diverse range of societies, empires, and civilizations, not to mention Snapchat and selfie sticks, which provided new elements to express our humanity.[1]

Though in the grand scheme of things, AI is just humble computer code designed to make human tasks more predictable, history teaches us that even mundane technological innovations can have big psychological consequences when they scale. Consider the way major human inventions have rewritten our modus operandi in the absence of any major biological changes. As Will and Ariel Durant write in *The Lessons of History*: "Evolution in man during recorded time has been social rather than biological. It has proceeded not by heritable variations in the species, but mostly by economic, political, intellectual, and moral innovation transmitted to individuals and generations by imitation, custom, or education."[2] For example:

- *Economic:* the stock market, derivatives trading, the gig economy, and non-fungible tokens (NFTs)[3]

- *Political:* communism, fascism, liberal democracies, and state capitalism

- *Intellectual:* relativity theory, Bach's *Well-Tempered Clavier,* the Google search engine, and Shazam

- *Moral:* every religion in the world, humanism, our family values, and our self-righteous minds[4]

Where we go from here is up to us. As the AI age progresses, we have to find new ways to express our humanity—for good or bad or both.

So far, the most consequential aspect of AI is not its ability to replicate or surpass but rather *impact* human intelligence. This is happening, not through AI's inherent capabilities, but through the digital ecosystem we have built to harvest, refine, and deploy AI at scale. This ecosystem, which may be to the metaverse what dial-up internet was to Wi-Fi, has positioned AI as a ubiquitous and powerful influence on human behavior. Like any powerful force, there will be positive and negative consequences for social behavior. But change is the salient feature and what makes the AI age a significant phase in our human evolution. This change has three main enablers, namely, a hyper-connected world, the datafication of you, and the lucrative business of prediction. In the next sections, I discuss these enablers in more detail.

A Hyper-Connected World

To say that we inhabit a hyper-connected world is as big a cliché as to say that the present is unprecedented and the future uncertain, or that a company's greatest asset is its people, which, alas, never stops anyone from offering these platitudes anyway. And yet, it is still true that the world has never been as connected as it is today.[5] Hyper-connectedness is one of the defining features of our time.

We are living much more connected lives than we ever have, and the trend is still upward.[6] It has never been harder to be isolated from other people and information—true facts and fake news—at least without checking into a meditation retreat. Never before has it been so easy to communicate with strangers, make new acquaintances, turn strangers into instant dates or future spouses, and maintain deep psychological contact with people regardless of who they are, where we are, and whether we've ever met them.

As much as technology has taken over our lives, and as much as we're hyper-connected, at the same time our current behaviors are simply catering to our preexisting human desires. From a psychological perspective, things haven't actually changed that much.

For instance, every time we hit "refresh," we are trying to validate ourselves, monitor our reputation, or answer a deep psychological question about our existence and the meaning of life: for example, what is going on, what do people think of me, what are my friends up to, and how am I doing in life? Our early predecessors shared these fundamental questions thousands of years ago; the only difference is they didn't have smartphones or the luxury of devoting so much time to engaging in these self-obsessed, neurotic ruminations.

If an average human from the 1950s were transported to our present times, *Back to the Future*–style, what would she see? Unlike Marty McFly, she would not see bionic X-ray vision implants or self-tying shoes, but simply wonder about the fact that most of us are glued to our mobile devices, irrespective of knowing whether algorithms are quietly working their magic in the background, or that we are engaging in unprecedented degrees of inappropriate self-disclosure, sharing our unsolicited views and news on everything and anything with everyone and anyone, for no other obvious reason than the fact that we can.[7]

Our visitor from the past may arguably be disappointed. In the famous words of contrarian entrepreneur Peter Thiel, "We were promised flying cars, instead we got 140 characters."[8] Though much of what this time traveler would see would be new, I doubt she'd have much trouble adapting to our way of life. Give her a smartphone, show her how it works, and everything will work out fine.

What keeps us so utterly immersed in the hyper-connected digital universe we created, and the very reason this universe exists in the first place, is our deep desire to connect with one another, which caters to our primordial needs.[9] The foundations of our hyper-connected world are largely the same universal needs that have always underpinned the main grammar of human life.[10] The need to *relate* to others, the need to *compete* with others, and the need to find *meaning* or make sense of the world. These three basic needs can help us understand the main motives for using AI in everyday life.

First, AI fulfills our *relatedness* need, that is, the desire to connect and get along with others, widening and deepening our relationships and staying in touch with friends. There's a reason we refer to social media platforms as "social networks," a term that has always been used to describe the web of friends, contacts, and connections we have, representing our basic social capital.

Second, AI can be seen as an attempt to boost our productivity and efficiency, and improve our living standards, all of which address our need for *competitiveness*. To be sure, we can (and should) examine whether this has been accomplished or not, but the intention is always there: to achieve more with less, increase work output and efficiency, and most obviously to increase consumption—the accumulation of resources.

Third, AI is also deployed to find *meaning*, translating information into insights, helping us to make sense of an otherwise ambiguous and complex world. For better or worse, most of the facts, opinions, and knowledge we access today have been curated, organized, and filtered by AI, which is why AI can be equally powerful for informing or misinforming us.

The big players of the AI age have created virtual platforms where we can express and fulfill our universal needs. Take Facebook, LinkedIn, TikTok, or any popular social media app. These platforms have become the main habitat of AI because they can connect us with others (*relatedness*), creating a level of on-demand psychological proximity to people's personal and public lives, irrespective of our real closeness to them. They also enable us to show off, advance our careers, display our virtues and status, and showcase our levels of confidence, competence, and success (*competitiveness*). Equally important, though perhaps less obviously, we can use the major social media apps to fulfill our insatiable appetite for sense-making (*meaning*), helping us find out who does what, when, and why, within the ever-expanding orbit of people's public reputation and their ever-shrinking private life.[11]

As decades of scientific research suggest, we are all "naive psychologists," or amateur explorers of humanity, looking to make sense of others' behaviors.[12] One of the consequences of being a hyper-social and group-oriented species is an obsession with understanding or at least trying to interpret what people do and why. Whether we realize it or not, this obsession has fueled the vast application of AI to social networking platforms.

These deep psychological functions of our hyper-connected world became crystal clear during the Covid-19 pandemic, which highlighted technology's power to keep us productive, as well as socially and emotionally connected, even in extreme physical isolation.[13] For much of the industrialized world, and especially for knowledge workers, aside from the nontrivial issue of staying healthy and sane, not to mentioned married, the much discussed "new normal" differed from the old normal only in degree; we basically increased our already high screen time.

So, we used Zoom to work and drink with friends and forgot why the office exists.[14] Likewise, at a time of much uncertainty and confusion, our digital hyper-connectedness gave us tools to access knowledge (e.g., Google, Wikipedia, Udacity, and WikiLeaks), systems of meaning (e.g., religious and political groups, Fox, CNN, the Johns Hopkins Hospital Covid-19 microsite, and Netflix), as well as access to infinite music libraries and podcasts, self-proclaimed and real experts on everything pandemic related, and every major literary work in the world.

But some of this is problematic. How much time are we spending not *truly* connected to other things or people, in the analogue or real sense of the word? Not much. We have turned ourselves into human wearables, attached to our phones nonstop, with additional sensors from our smart watches, Oura Rings, Siri, and Alexa, while we patiently await to upload our memories, fantasies, and consciousness to the cloud. In a relatively short time frame, we quickly transitioned from the internet to the internet of things and now the "You of Things," a concept that sees our bodies as part of an enormous sentient digital network, and our entire existence downgraded to the status of our smart TVs and refrigerator.[15] Since our selves have been largely reduced to the digital fragments of our reputation captured in the many devices that connect us to others and the world, it is hard to disagree with Yuval Harari's premise that "we are becoming tiny chips inside a giant data-processing system that nobody really understands."[16]

Some say that AI has turned humans into the product of tech firms, but a more accurate description, as Nobel Prize–winning novelist Kazuo Ishiguro recently noted, is that we are more like land or soil being harvested or excavated, with the real product being data, and its value being based on the ability to influence or change our beliefs, emotions, and behaviors.[17]

The major change from twenty years ago is, arguably, the amount of data we have produced and continue to produce to the point of translating every possible human behavior into a digital signal of it. We are, more than ever, not just physical but also virtual creatures, and our existence has acquired a second life in the form of virtual records encoded in the cloud, stored in ginormous data warehouses.[18]

The behavioral DNA of our habits, including our most intimate preferences; our deepest, most private thoughts; and our guilty pleasures, has been turned into a vast reserve of information so that the algorithms can learn all there is to know about us. Scientific studies show, unsurprisingly, that AI can make more accurate estimates of our personality than not just our friends but also ourselves.[19]

The Datafication of You

Our impetus to understand and predict the world, including ourselves and other people, underpins much of the current AI age, which is founded on the premise and promise of gathering as much data on people as possible, turning us all into the subjects of a massive psychological experiment.[20]

When I was running research experiments for my doctoral thesis, barely twenty years ago, I had to drag people into the testing cubicle and beg them to complete a psychological assessment. Gathering data from fifty people could take months, even if we had the funds to pay them. Today we have more data on humans and every single aspect of our behavior than we could possibly analyze. We could stop collecting data and spend the next century trying to make sense of it and still barely scratch the surface. Almost everything we do creates a repository of digital signals representing the fuel or gasoline that enables AI's intellectual development.

To be clear, more data doesn't make people more predictable: data is just a record of what we do; it is the product rather than the cause of our activities and behaviors. However, the very platforms and tools that are deployed to get us to produce ever more data do a great job of standardizing our main patterns of activities, incentivizing us to act in more predictable and repetitive ways. Consider how Facebook, a platform that actually allows for a relatively rich repertoire of interactions and variety of interpersonal activities, constrains the range of responses or behaviors we can display.

Sure, we can express our comments in the form of unstructured—even creative—text. But it's much easier to like, share, or insert emojis in response to what we see so that we can focus our energies on labeling the people in our photos, tagging others in stories, and encoding the rich variety of information into highly structured, standardized data, which provides clear-cut instructions for AI. We become the unpaid supervisors of machine learning algorithms, as well as their object of study, albeit in a simplified, repetitive form.

Unsurprisingly, dozens of scientific studies indicate that Facebook likes and other forced-choice categories of expression accurately predict our personality and values.[21] Think of likes as the digital equivalent of bumper stickers, rebellious teenagers' T-shirts, or tattoos: humans are proud of their identity, so they relish any opportunity to share their views, beliefs, and opinions with the world, in part to delineate an in-group and out-group. It is not prejudiced, but socially insightful, to assume that a car with a "hipster killer" bumper sticker has a very different driver from one with a "keep calm and go vegan" sticker.

This is even more obvious with Twitter, where the variety of input data (the content and context of tweets) can be consistently mined to predict retweets, irrespective of whether we read or processed the information. The platform introduced a "read before you retweet" feature to encourage responsible sharing: Perhaps the next feature could be "think before you write"?[22] If Twitter's algorithms are often accused of augmenting our echo chamber, that's only because they are trained to predict what we prefer to attend to—stuff congruent with our views and beliefs.[23] In essence, we turn us into a more exaggerated version of ourselves, not open-minded but narrow-minded.

Even among the tech giants, Facebook stands out for its single-minded bet on data, which explains why in 2014 it splashed out $19 billion to acquire WhatsApp, which, at the time, had just fifty-five employees, no more than $10 million in revenues, $138 million in losses, and a valuation of $1.5 billion just one year prior.[24] As Larry Summers pointed out during a seminar at the Rotman School of Management, "Everything WhatsApp had, all the people, all the computers, all the ideas, could fit in this seminar room, and there would still be room for several seminars."[25]

While this provides a clear-cut picture of the new realities of the digital economy, Summers forgot to mention that the most valuable thing WhatsApp owned did *not* fit in that seminar room, namely, the vast amounts of data and all the users worldwide committed to producing more and more of it every day. When WhatsApp was acquired, it boasted 450 million users. Today, the number exceeds 2 billion. Facebook itself has 2.8 billion users who spend around two hours and twenty-four minutes on the platform every day, with an extra thirty minutes spent on WhatsApp, which boasts 60 percent of all global internet users and is the number one messaging app in 180 of the world's 195 countries.[26]

In 2021, Facebook—now Meta—took the datafication of you one step further by merging WhatsApp and Facebook data to deepen its knowledge of users. This is the power of combining everything you do on the number one social media platform with everything you say on the number one messaging and free call app—oh, and there's your Instagram footprint, too. Likewise, the datafication of you has enabled Netflix to go from binge-worthy movie recommendations to blockbuster content creation, and allowed Spotify's AI to teach artists how to create more popular songs, sharing its vast consumer insights with them and educating them on what their actual and potential audiences like and dislike.[27] In a not-so-distant future, AI advances in musical composition may enable Spotify to automate some of its artists, just as self-driving cars would enable Uber to automate its drivers. Uber drivers currently have two jobs: taking customers from A to B (the official job), and teaching AI how to do this without human drivers (the unofficial job, which justifies the valuation of a loss-making company at $24 billion). In a similar vein, imagine a world in which the platform's AI learns to *create* (not just curate) music in direct response to your preferences, turning Ariana Grande and Justin Bieber, the two most popular artists on that platform, into musical relics (I will leave it up to you to decide whether these hypothetical technological advances would represent a form of artistic progress or not).

While many of the services provided by the big and not-so-big AI companies are free, in the sense that we don't pay for them with money, investors value them because of the perceived value assigned to the data the companies ingest, analyze, and sell.[28]

Fundamentally, our digital records have enabled tech firms to persuade others—in particular, financial analysts, investors, and the market—that they have an accurate understanding of us, including our unique selves, which explains the exorbitant valuations of data-rich firms and any business convincingly claiming to be in the lucrative business of using AI to predict human behavior.[29]

The Lucrative Business of Prediction

AI has been sensibly described as a prediction machine, since algorithms demonstrate their "intelligence" by forecasting things, which should in turn make our own decision-making more intelligent.[30] If data fuels the digital revolution, the value of data is based on its promise to decode human behavior, with a new level of granularity, scale, standardization, and automation. There has never been a higher dollar multiple paid for services capable of turning data into insights, all courtesy of AI. According to PWC, AI will contribute $15.7 trillion to the economy by 2030, increasing GDP by 26 percent.[31]

This new economic order is possible because of the combination of vast sets of big data and cheaper and faster computing power to crunch them and translate them into automated insights and nudges, and shaping human activity in commercially advantageous ways. For instance, Google's AI enables the company to convince clients that it knows their customers with sublime precision, which explains why 80 percent of Alphabet's revenues ($147 billion) still come from online advertising.[32] Likewise, the vast access to ubiquitous consumer behavior that Meta—Facebook's, Instagram's, and WhatsApp's parent company—has harnessed enables the tech giant to leverage AI to sell extremely targeted content and personalized advertisement, customized to the world's wants, wishes, and habits.[33]

It is, of course, also possible because we can't avoid spending so much of our lives online, and because of a critical feature of humans: though we hate to admit it, we act in consistent and predictable ways, to the point that there are clear identifiable patterns underlying our unique habits and everyday behaviors—a sort of personal *syntax of you*. This syntax is precisely what AI monetizes: every thought, value, and idea recorded, the stuff that makes you you and distinctively different from others. Just as you could work out many things about a stranger by looking at the history of their browser (unless they deleted it, which would be a revealing data point in itself), the algorithms that mine our lives are quite good at predicting what we might do next—and they are getting better. When, just ten years ago, Target's AI determined that a woman was pregnant before she had even decided to share the news with friends and family (all based on her shopping patterns), it all seemed like a creepy episode of *Black Mirror*.[34] We are now well aware of what algorithms know or may know about ourselves and others; when it comes to AI, creepy is the new normal.

By analyzing our every move and trading its insights on how to influence us at a high price to brands and marketers, AI is in effect selling human futures, attaching new value to the "behavioral surplus" it derives from all the data we generate. And while all this is perhaps justifiable through our choices and preferences— fast, cheap, predictable, and efficient optimizations of our everyday needs—it is a pity that we are arguably becoming less interesting and creative in the process. Even if AI's goal were not to automate us, it appears to be turning us into automatons.[35] So far, our data is predominantly commercialized for marketing purposes, such as targeted ads, but there has already been a range of incursions into many other areas, such as life insurance, career success, health and well-being, and romantic relations. For example, China uses AI to translate mass behavioral surveillance into a credit score and, in turn, a management system for its citizens.[36] Imagine you are rude to a taxi driver, you forget to tip the waiter or to cancel a restaurant reservation, or you drive through a red light, and any of these actions automatically reduces your ability to secure a mortgage, credit card, or job.

The lucrative business of prediction has also permeated the realm of love. Consider Match Group, which owns a percentage of many of the world's most popular dating sites, including Tinder, OKCupid, Hinge, Plenty of Fish, and Match.com.[37] Its chatbot Lara interacts with its global users to collect as much personal data on their romantic relationship preferences as possible, which in turn allows users to consume the ads that fund their digital love expeditions, especially if they are interested in avoiding a paid subscription. Or LinkedIn, which sells monthly membership services to recruiters so that they can access data on the skills, résumés, and background of candidates that are *not* in their personal network. This is information it gets for free, because LinkedIn members volunteer it, in part to get a better job (LinkedIn estimates that 70 percent of its 775 million members are at least open to this), if not to attract clients, impress friends and colleagues, consume curated media, or just follow news stories.

In a remarkable book, Shoshana Zuboff refers to the lucrative business of prediction as "surveillance capitalism," "a new economic order that claims human experience as free raw material for hidden commercial practices of extraction, prediction, and sales" as well as "a parasitic economic logic in which the production of goods and services is subordinated to a new global architecture of behavioral modification."[38] Zuboff's poignant critique of the AI age explains why people fear the power of Big Tech, and why documentaries like *The Social Dilemma*, in which former Facebook employees come clean about the cynical Machiavellian tactics of manipulations behind the platform's algorithms—from addictive game-like features to psychological nudges to decode and mold users' behavior—are rather shocking to many.

The mere fact that you may not be experiencing life in this Orwellian way highlights the immersive allure of the system itself, which has managed to camouflage itself as a normal way of life, successfully turning us into a rich record of digital transactions immortalized for AI's posterity. A fish doesn't know what water is; same goes for humans and the matrix.

Thus, at least for now, AI's influence is not so much a function of either emulating or surpassing human intelligence, but shaping the way we think, learn, and make decisions. In this way, AI is molding the very object it attempts to recreate, like a great master tinkering with an object she is about to paint. If you want to copy a drawing and you have the ability to simplify the model in order to draw a closer replica, it makes the task easier.

Most of us aren't scientists, yet in any area of life we ordinarily operate according to AI's core principles: using past data to not just predict and decide on the future. When we buy a product Amazon has recommended to us, watch a movie Netflix has suggested, or listen to a playlist Spotify has curated for us, we are making data-driven changes to our life, conforming to an algorithmic syntax that eliminates the behavioral differences between us and people like us, boosting the valuation of tech firms by making our lives more predictable. Prediction improves through two different ways: the algorithms get smarter or the humans get "dumber." The latter implies that our ability to respond to a situation in different ways, control our reactions to stimuli, or *own* our behaviors in an agentic, self-controlled way decreases. Every minute we spend online is designed to standardize our behaviors and make us more predictable.

The initial wave of the Covid-19 pandemic saw physical connectedness instantly replaced by digital hyper-connectedness, with Big Tech firms reaping the benefits. During 2020 alone, the market cap of the seven biggest tech firms, which include Apple, Microsoft, Amazon, and Facebook, increased by $3.4 trillion.[39] As shops worldwide closed, Amazon's earnings rose by 40 percent in a single quarter alone, and its Web Services, the world's biggest cloud platform, saw similar increases, as an unprecedented number of brick-and-mortar businesses were forced to go virtual.[40]

In the United States, spending on online shopping during the pandemic increased by 44 percent during 2020, producing more growth in a year than during the entire prior decade.[41] Of course, the growth of online retail is intuitive, as shops were forced to shut down and we consumers were left only with the choice to buy online. But an even bigger switch was to shut down in-person interactions altogether, driving all forms of contact, communication, and exchanges to the cloud. The idea of a metaverse, a shared virtual space that takes over our physical world, went from far-fetched digital dystopia to inevitable and imminent reality in less than two years. While many people died, got sick, or lost their jobs or their entire businesses, and most businesses saw significant hits due to the pandemic, many more people merely increased their online consumption, making the rich tech firms substantially richer, including in terms of data.

As the *Economist* reports, the use of data is now the world's biggest business.[42] In May 2021, the share of the S&P accounted for by the five biggest tech juggernauts—Apple, Amazon, Microsoft, Alphabet, and Facebook—had increased to almost 25 percent, from 15.8 percent a year earlier.[43] This takes their combined market cap to over $8 trillion, which is more than the smallest three hundred S&P firms put together.[44] To put this into perspective, the GDP of the biggest economies in the world (in trillions) is the United States, $20.5; China, $13.4; Japan, $4.9; Germany, $4; and UK, $2.8. The astronomical valuations of Big Tech are predominantly based on the predictive data these companies have managed to harness. Your probability of having given no data to any of these firms is about 0 percent. By exactly the same logic, as lockdowns eased and the pandemic got under control, much of the world returned to offline or analogue activities, shedding over $1 trillion from the valuation of Big Tech firms.

Although these valuations rest on the belief that AI enables the data-rich tech firms to understand us better, and that this deeper level of understanding can help them change, influence, and manipulate human behavior at scale, AI's ability to truly understand us has been exaggerated. This is why the ads we are shown through data-driven algorithmic targeting rarely produce a *wow* effect, and generally fail to create mind-blowing insights about our deepest consumer preferences; they seem either creepy or crappy, like when they show us a pair of sneakers we've already bought or a vacation hotel we decided not to book. But we are still exposed to them and keep consuming the stuff they show us anyway.

So far, the main accomplishment of AI has been to reduce some of the uncertainties of everyday life, making things—including ourselves—less unpredictable, and conveying a sense of certainty in areas that were always seen as serendipitous. Each time we spontaneously react to AI, or one of its many manifestations, we do our bit to advance not just the predictive accuracy of AI, but the sterilization of humanity, making our species more formulaic.

Our Dark Side Unleashed

Two of the most famous Enlightenment philosophers, Jean-Jacques Rousseau and Thomas Hobbes, wondered whether humans are either born good and made bad by civilization, or rather born rotten but "civilized" by it.[45] The question at stake here: Is human nature good but society destroys it (Rousseau), or are humans quite useless and immoral to begin with, which society tackles or remedies by somehow taming us (Hobbes)?

The answer, as with most either-or questions pertaining human behavior, is yes, or "a bit of both." Humans are unique and complex creatures, so we interact with society—including technology—in a wide range of ways. Our relationship with AI is no different. AI is at times a magnifier, and at times a suppressor, of our own character, dispositions, and nature. And the digital echo-systems in which we coexist with AI, such as social media platforms, enable us to express our cultural identity, norms, and traditions.[46]

In our interactions with AI, we can see psychological patterns emerging as cultural bastions of our behavioral DNA, which may at most contrast in degree to what we would typically do in past eras. Equally, we can see our seemingly new AI-infused habits eroding some of the cultural differences that have always defined normative behaviors—"how we do things around here"—between different cultures and societies, or between different periods

within the same culture. Culture can comprise any socially transmitted blueprint or etiquette that makes one group of humans
(e.g., Starbucks employees, Canadian citizens, Portland hipsters,
and Hasidic Jews from Williamsburg) unique, or at least different
from others (e.g., IBM managers, illegal aliens, nineties yuppies,
and the American-Taiwanese community in Los Angeles). So, for
example, Italians are generally more extraverted and sociable than
Finns, but this is less evident when Italians and Finns use social
media, which operates as a digital suppressor of cultural heritage,
prompting everyone, including Finns, to share their unsolicited
thoughts, likes, and emotions with the rest of humanity, as if they
were Italian, even if the outcome is that everyone ends up living
their lives like introverted computer nerds. Frank Rose, the former
editor of *Wired*, noted a decade ago that our current world is basically a footnote to 1980s *Otaku* culture in Japan, the subculture in
which teens escape the real world to live in a universe of fantasized
manga or anime characters and gamified relations with fictional
others.[47]

The adaptive and the maladaptive, the virtue and the vice, depend not so much on universal systems of value, or on subjective
moral conventions, but on their effects on ourselves and the rest of
humanity at a given point in time. Every habit or behavioral pattern has been within us, in our rich range of behavioral repertoires,
since the dawn of humankind.[48] But what we express, and whether
this is deemed good or bad, can be judged only relative to its individual and collective outcomes. As Will Durant noted, "Every vice
was once a virtue, and may become respectable again, just as ha-
tred becomes respectable in wartime."[49] There is no way to judge
ourselves but with ambivalence, accepting the ambiguity of human
behavior and the complexity of human nature.

Most of the problematic patterns we may condemn during our times—from the sedentary habits and excess of fast-food consumption that trigger obesity to the ADHD-like behaviors caused by our compulsive smartphone addiction and excessive screen time—are likely caused by a mismatch between ancient human adaptations and current challenges, which render those ancient adaptations obsolete, if not counterproductive.[50]

For instance, greed was probably a virtue in times of extreme food scarcity, focusing the human mind on the ruthless accumulation of resources and optimizing life for sheer survival.[51] But when food is plentiful and resources are abundant, it is self-restraint rather than greed that turns into a virtue, and greed that carries the seeds of self-destruction. By the same token, curiosity may be a virtue in groups or societies that put a premium on learning, general knowledge, and open-mindedness, but a curse if the desire to explore different environments, places, or people may more likely endanger us or distract us to the point of deteriorating our focus and productivity.[52]

The dark elements of human behavior are whatever we consider undesirable, toxic, counterproductive, or antisocial in the face of specific adaptational challenges posed by our current environment. Simply put, the dark side of AI is the dark side of humans in the AI age, because AI, like any influential new technology, has the power to not just reveal but also amplify undesirable human qualities, such as our impulsive, distractive, self-deceived, narcissistic, predictable, lazy, or biased nature. When we blame AI, or indeed any novel technology, for dumbing us down, corrupting us, or turning us into seemingly obnoxious or unpleasant creatures, what's revealed is a disconnect between historically adaptive tendencies or predispositions and new environmental challenges—today's major challenge is AI.

This book is about how AI has not only exposed but also augmented some of our worst character traits. Think of this as the human sins of the preautomation age. If we want to reclaim our humanity and remind ourselves, as well as any potential visitors from Mars (assuming they can afford the ride with Space X), of our alleged special status as a species, then we must learn to control our maladaptive tendencies and rediscover the qualities that make us special.

In that sense, the most notable thing about AI is not AI itself, let alone its "intelligence," but its capacity for reshaping how we live, particularly through its ability to exacerbate certain human behaviors, turning them into undesirable or problematic tendencies. Irrespective of the pace of technological advancement, and how rapidly machines may be acquiring something akin to intelligence, we are as a species exhibiting some of our least desirable character traits, even according to our own low standards. This aspect of the AI age ought to concern us most: this is not about automating humans, but degenerating or deteriorating humanity.

Chapter 2

Weapons of Mass Distraction

How the AI age turned life into a big interruption

Where your attention goes, your time goes.

—Idowu Koyenikan

5:03 p.m., October 7, 2020

I am sitting in my Brooklyn home office in front of three screens, each of which has at least seven apps open. My desire to avoid human, or shall we say analogue, distractions in the house (i.e., lockdown kids) is why I'm wearing noise-canceling headphones, which are playing my Spotify songs and enabling me to ignore any background noises. The people on my Zoom call, however, aren't so lucky. Because of my high-fidelity microphone, which is extremely sensitive to sound, they are acoustically closer to my children and my domestic mayhem than I am.

As I write, I add different songs to my playlist, which I also share with friends. They are commenting on my songs over WhatsApp and sharing their own music. In the process, we catch up on world and soccer news, as well as some low-quality gossip. Meanwhile, messages arrive with their characteristically irritating alert sounds to my different email addresses, and I alternate between my ever-expanding to-do list and my chaotic calendar, which reminds me of my early wake-up time tomorrow (I am speaking at a virtual conference in Singapore at 4:30 a.m. my time). With all these exchanges and distractions, I feel like a physical vessel that generates digital records of its unfocused schizotypal mind, translating brain activity into a codified series of 0s and 1s to nourish AI's insatiable appetite.

●　●　●

It is now 10:36 a.m., December 29, 2021. I am in Rome, with the same computer, devices, and an even bigger range of apps colliding to interrupt my impoverished attention. As you may guess, it's a miracle that this book was ever finished. Thankfully, I'm not writing a book as long as *In Search of Lost Time* (4,215 pages), though Proust's title is rather appropriate for the age we're living in.

The Two-Second-Attention Economy

While the deployment of AI as a distraction tool is a relatively novel phenomenon, the economics of attention and the commoditization of our interest and preferences has a much longer history.

Decades ago, the psychologist and Nobel laureate Herbert Simon first pointed out that humans were struggling with information overload, and that "the wealth of information" we encounter every day creates a "poverty of attention."[1] We tend to value things more when they are scarce, and an overabundance of anything will commoditize and trivialize that thing—for example, cereal brands, TV channels, and unsolicited emails flooding your spam folder with "great opportunities."

Though the attention economy was kick-started by the invention of the printing press in the nineteenth century, which radically expanded the spread of information as publishers competed for readers' time and attention, computers took it to a whole other level.[2] Simon predicted that as information, including the technologies to record and disseminate it, grows exponentially, the competition for human attention, a finite resource, will intensify even more: "What information consumes is rather obvious: it consumes the attention of its recipients."[3] His words, written fifty years ago, have become famously prophetic. When we all trade information, it is not just information that becomes devalued but also our attention. I recall, during my first visit to Tokyo nearly twenty years ago, the sensory bombardment from spending just a few minutes in the underground—every inch of space was taken by an ad, constant announcements, and commuters playing games or watching movies on their devices—or in the emblematic pachinko parlors. Fast-forward to today and we are all immersed in a similar sea of sensory stimulation pretty much anywhere.

The battle for focus—for mere seconds of our attention—has reached epic levels and has been intensified by data-driven metrics such as clicks, likes, views, and tags, which are critical to improve AI's ability to understand and influence consumers. Attention and data are the two key ingredients of the AI age.[4] Without attention, there is no data, and without data, there is no AI.

The quantification of our attention—and AI's ability to weaponize that information—creates a vicious cycle: since our attention is scarce and information is plentiful, the battle for our attention exacerbates. Netflix is competing with Twitter, Twitter is competing with the *New York Times*, and the *New York Times* is competing with Instagram; they're all competing for our precious time and our even more precious focus. Their algorithms crave our attention, and their business models depend on it, which makes our attention highly valuable, not least because there is so little of it left to capture after the algorithms consume it. It also leads to yet more information overload, which threatens to distract us even more.

The result is a degradation of focus that causes attention deficit hyperactivity disorder (ADHD)–like behaviors, such as restless hyperactivity, rapid boredom, and impulsivity (more on this in the next chapter). These symptoms are best evidenced during digital withdrawal: those twenty minutes on the subway or the six hours of downtime while flying over an ocean, which, unsurprisingly, have become rather optional in recent years, as more passengers hook onto in-flight Wi-Fi. If the internet, social media, and AI can be described as a distraction machine, the assumption is that whenever we attend to the contents of these technological attractions, we are prone to ignore life: it is arguably more appropriate to see life as the actual distraction, since, statistically speaking, it has been downgraded to an occasional psychological interruption from our nearly perpetual state of digital focus and flow.

At least 60 percent of the world is now online, and in developed nations, the average internet user will spend around 40 percent of their awake time being *active* online.[5] However, that *passive* online time—remaining connected to our devices and emitting data, even while we are not actively interacting with them—makes up a significant chunk of our remaining awake time. You need to be far away from your phone, computer, and wearables to truly be offline or live a purely offline existence. In my case, this does not even happen while I sleep, unless my smart watch and smart ring run out of battery power. The *un*quantified self is now more elusive than our quantified self, and in the rare event that our attention is truly devoted to something that cannot be found online, we will likely create a record of it online, as if the analogue life were not worth living. In 2019, Apple sold more watches than the entire Swiss watch industry combined, and workplace tracking software, which has increased dramatically with hybrid and remote working modes, is expected to be adopted by 70 percent of large firms in the next three years.[6]

The AI age has been described as "the most standardized and most centralized form of attentional control in human history. The attention economy incentivizes the design of technologies that grab our attention. In so doing, it privileges our impulses over our intentions."[7] It is as if we have been hypnotized by AI, possessed by a never-ending flow of information, immersed in a deep sea of digital distractions. More than ever before, we can be mentally absent and separated from our physical existence, which renders the metaverse far less futuristic than we may have thought. Our offline experiences and, indeed, existence are now very much out of focus. Inevitably, this will impact our ability to think seriously about important social and political issues, as if our brains were intellectually sedated by AI. As author Johann Hari points out: "A world full of attention-deprived citizens alternating between Twitter and Snapchat will be a world of cascading crises where we can't get a handle on any of them."[8]

A Life Less Focused

Although it is too soon to observe any significant effects of technology on our brains, it is plausible that long-term effects will occur. As Nicholas Carr, who writes about the intersection between technology and culture, noted in *The Shallows: What the Internet Is Doing to Our Brains*, repeated exposure to online media demands a cognitive change from deeper intellectual processing, such as focused and critical thinking, to fast autopilot processes, such as skimming and scanning, shifting neural activity from the hippocampus (the area of the brain involved in deep thinking) to the prefrontal cortex (the part of the brain engaged in rapid, subconscious transactions).[9] In other words, we are trading speed for accuracy and prioritizing impulsive decision-making over deliberate judgment. In Carr's words: "The internet is an interruption system. It seizes our attention only to scramble it."[10]

Some evidence, albeit preliminary, does suggest that certain tech-induced brain influences can be seen (and measured) already, such as white-matter changes in preschool children as a result of extensive screen use. An estimated 62 percent of school students now use social media during class, and up to 50 percent of classroom distractions have been attributed to social media.[11] College students spend a staggering eight to ten hours *per day* on social media sites, and as one would expect, time spent online is inversely correlated with academic performance.[12] Unsurprisingly, there is consistent research evidence linking higher levels of Facebook use with higher levels of academic distraction, which in turn lowers academic achievement.[13]

The notifications, messages, posts, likes, and other AI-fueled feedback rewards hijack our attention and create a constant state of hyper-alertness, interruption, and distraction, capable of generating significant levels of anxiety, stress, and withdrawal.[14] This is particularly problematic in young people, who are in the midst of their intellectual and identity development and depend on validation and feedback signals from others. Moreover, when our attention is co-opted by AI-enabled distractions, we tend to rely on more intuitive or heuristic decision-making, which includes triggering our biases, stereotypes, and prejudices, all of which are prone to making young (and not-so-young) people more narrow-minded and less inclusive.[15] To have an open mind, you need to be willing to proactively seek information that runs counter to your own attitudes, which is much harder—and less likely—when you are not paying attention and are at the mercy of AI's algorithms.[16]

Compelling scientific evidence indicates that distracting young people from social media tends to induce stress in them, much like separating smokers or drinkers from their addictive substances.[17] Indeed, lower levels of attention control have been associated with heightened levels of anxiety, and the wide range of digital distractions delivered by the AI age are a threat to our attention control because they monopolize our attention. The more vulnerable you are to attention-control deficits, the more your learning will be impaired by digital distractions. Thus, in people who are naturally prone to inattention because they have poorer potential for attention control, social media use causes significantly higher levels of psychological distress.[18] Being digitally hooked also impairs people's health and physical well-being. Already during the early stages of the AI and social media phase, academic studies reported a strong correlation between screen use and body mass index, and sedentary behavior has increased because naturalistic aspects of physical activity have been steadily decreasing over the past twenty-five years as internet technologies have become more popular.[19]

Professor David Meyer, a leading multitasking scholar, compares the damage of the AI age to the glory days of the tobacco industry: "People aren't aware what's happening to their mental processes in the same way that people years ago couldn't look into their lungs and see the residual deposits."[20] Although this may be an overstatement, it is clear that our typical patterns of focus have changed dramatically merely in the past fifteen years. To borrow the words of tech writer Linda Stone, we are living in an age of "continuous partial attention."[21]

Cognitive psychology has studied floating attention for decades. One of the most prominent theories of attention gave rise to the famous *cocktail party* effect, which explains the common experience of chatting to someone at a party until our attention is interrupted by the sound of a familiar word—perhaps our name or the name of someone we care about—spoken in the background by another guest.[22] Typically, we will then turn toward them and realize that we were actually also listening to them, at least in part.

This raises an interesting question, namely, whether our attention to the offline world is less focused than we think. Perhaps some things are on stage like in the theater, but there's stuff happening behind the scenes that we are not fully oblivious to, a bit like people whispering to each other in the cinema when you are trying to watch a movie. For instance, we may be having dinner with someone, trying to listen to a colleague during an in-person meeting, or playing with our kids, all without giving them our undivided attention. To achieve this, the AI age provides an endless flurry of automated distractions to fill up not just our mental stage but also our back stage, penetrating our offline experience and existence by keeping a part of us always focused on, if not perpetually dependent on, digital updates.

It would be nice to have a simple solution to these problems, but there isn't one. Technophobia may sound like a tempting option, but it comes at a high social and economic price, namely, turning us into useless and unproductive citizens of the AI age, where Luddites are rarely accepted or integrated. Being offline equates to having your entire existence ignored, like the mythical tree in the forest that falls when there is nobody there to hear it. Blocking apps or restricting internet access is an obvious in-between compromise, allowing us to refrain from at least some digital distractions.[23]

Unlike our evolutionary ancestors, who were rewarded for absorbing as much of their sensory surroundings as they possibly could, what's adaptive today is the ability to ignore rather than carefully surveil our environments, because we inhabit a world of noise, distraction, and ubiquitous sensory garbage. In times of information overload and nonstop digital media bombardment, distraction is destruction of our precious mental resources, and the only recipe for productivity is ruthless self-discipline and self-control. Fittingly, the internet is plagued with self-help advice on how to avoid distractions and boost our focus and productivity, all of which are inversely related to the time you waste internalizing this online advice.

When legendary jazz genius John Coltrane, going through a compulsive saxophone-playing phase, lamented to fellow band member Miles Davis that he was unable to stop playing the sax, Miles replied, "Have you tried putting the damn thing out of your mouth?" We could apply the same logic to controlling our most recurrent digital distractions. For example, I found that a useful way to avoid being distracted by Facebook is to delete the app altogether. Who would have thought? Granted, there are times when I miss the nostalgic sentiment of snooping on my high school friends or monitoring what strangers are having for brunch or seeing whether my distant cousin has checked into the business lounge, but I somehow suspect this may not be a terrible loss for civilization.

Unproductivity Tools

Most technologies are designed with the purpose of improving our productivity. However, while the early wave of the digital revolution—late 1990s and early 2000s—did register relative productivity increases, this can largely be attributed to the creation of IT-intensive industries and the related expansion of the knowledge economy. Basically, a whole new sector was created, which enabled lots of people to upskill and tons of new businesses to emerge. Imagine that everything that ever existed offline had to migrate online, too. However, as the *Economist* notes, "Since the mid 00's productivity growth has tumbled, perhaps because the burden of distraction has crossed some critical threshold."[24]

In other words, the same technologies that enabled us to make work more efficient through democratized and scalable information management tools, such as email, the internet, and smartphones, introduced a range of new distractions that disrupted our potential productivity gains. Some estimates suggest that workers reach for their smartphones for non-work-related activities as often as twice a minute, and that the task recovery time after a typical digital interruption—for example, checking your email, sports results, Facebook, or Twitter while you work—may be as high as twenty-three minutes.[25] If you still think we can become more productive, then the bar must have been very low to begin with.

Knowledge workers, who are more likely than other workers to spend their working days digitally connected, and who were once the main beneficiaries of digital technologies, waste an estimated 25 percent of their time dealing with digital distractions.[26] The *Economist* reckons that digital distractions cost the US economy—where knowledge workers account for at least 60 percent of GDP—as much as $650 billion every year.[27] Academic studies suggest that the productivity loss caused by distractions could be a staggering *fifteen* times higher than absenteeism, sickness, and health issues. Nearly 70 percent of workers report significant productivity deficits due to smartphone distractions.[28]

Multitasking is a nice idea but also a great myth. It is comforting to think that we can accomplish more by doing multiple things at once, and perhaps it even makes logical sense. People tend to perceive multitasking as an efficient productivity strategy, but research indicates that most of what they see as multitasking is in fact simply task switching.[29] Estimates indicate that multitasking deducts the equivalent of ten IQ points from our performance and is twice as debilitating as smoking weed (and, one would assume, less enjoyable).[30] There are actually some parallels, as seeing or hearing a smartphone elevates our levels of cortisol, just like marijuana does. Although smartphones can obviously be tremendously helpful to make us work more efficiently, prepandemic estimates indicated that between 60 percent and 85 percent of smartphone use occurring during working hours was devoted to nonwork activities.[31] Cyberloafing was once defined as the amount of worktime people spend checking the internet, but now many people appear to occasionally work during their internet time, inverting the previous imbalance.

Certainly, those who worry about whether people are able to work productively from home overestimate the percentage of actual time people devote to work when they are in an office, irrespective of whether distractions are caused by colleagues or smartphones.

It is somewhat ironic that the main reason most people tend to oppose hybrid work, in particular working from home, is the assumption that those who are not motivated enough to be productive at home will somehow want to be productive when they are in an office. Unless, of course, the office could provide a welcome distraction from 24-7 digital immersion, reminding people that work also allows in-person connections with others and that it is often quite stimulating and rewarding to interact with three-dimensional, shoe-wearing humans. In the AI age, being distracted by fellow office workers may be a welcome distraction from AI and a nostalgic breath of fresh air.

Interestingly, AI firms should worship working from home, since it means more attention and data from everyone, and stronger adoption of their tools (courtesy of the lower levels of physical or real-world interference), but when it comes to their own employees, they still prefer to bring people back to the office.[32] Scarface's "don't get high on your own supply" mantra is surprisingly relevant to the world of Big Tech.

Lost in the Search for Meaning

Self-help and productivity articles abound on how often we should use our devices, and the answer is always "less." In theory, we can all agree. In practice, we don't, which is why we never actually detach ourselves from our screens.

There are more fruitful ways to live our days than doom-scrolling on Twitter, and it is easy to spot the paradoxical nature underpinning the thriving market for apps allegedly devised to help us combat our smartphone addiction, like people who "reply all" to complain about people who hit "reply all."[33] However, these naturalistic fantasies of what is perhaps a more organic and mindful approach to life seem rather far-fetched for those who depend on technology to not just socialize, learn, and kill time but also try to be productive at work, as we learned when we all used technology to remain professionally active during the pandemic.

We did, after all, have the choice to spend our lockdown hours *not* binge-watching YouTube or Netflix, which increased paid subscribers by 31 percent in 2020, each watching an average 3.2 hours content per day.[34] Equally, nobody forced us to increase our use of Zoom or Microsoft Teams, as opposed to just using the phone, or to spend yet more time on Facebook and Instagram, as opposed to reading books, tidying our house, or baking bread, which did become a popular activity during the lockdown, at least based on people's social media updates. By the same token, the recent explosion of podcasts illustrates that attempts to escape the crowded digital space of ephemeral social media updates to focus our mind with more substantive content are quickly crowded and hijacked by content overload. Spotify went from 450,000 paid podcast subscribers in 2019 to 2.2 million in 2021 and now boasts as many podcasts.[35] Where to start? What to pick? How to stay focused on one, when there are so many other options?

As many parents will have realized by now, it is no longer rational to ask children to stop playing computer games or get off their iPads when the parents themselves are barely able to disconnect and the physical world provides no obvious alternative distractions.

Interestingly, Steve Jobs once noted that his kids were not allowed near his iPad and that they would never have one.[36] A few days ago, a parent I know expressed deep anger and indignation at the fact that her preschooler had stolen and hacked into her iPad to hide for hours inside a closet to watch videos. Judging by what parents do all day, it would be more logical for kids to rebel by secretly going out to play in the park while their parents force them to spend the day on their iPads. Likewise, we all get annoyed when our partners are constantly on the phone, while we are constantly on ours.

Although we have been lamenting our excessive screen time for a while, the trend is still upward, including for older people. A 2019 Nielsen survey found that Americans sixty-five and older spent ten hours a day or more on their screens. Even for those of us who average "only" around three to four hours of daily screen time, which is below the national average, that still means about ten years of our lives spent on a screen.[37] It's safe to assume that most of us feel guilty about these stats, though denial is always a soothing alternative. The question is, why don't we give up screen time for more real-world interests: our kids, school, the moonrise, the fallen leaves? The reason? Our online FOMO (fear of missing out) is bigger than any regrets that may arise from missing out on any real-life activities.

I still recall my sense of surprise when, during a marketing conference ten years ago, there was a growing interest in understanding the impact of our second or third screen, which alluded to the fact that people were starting to watch TV shows while browsing their iPads and perhaps also chatting to their friends over the phone about one or the other. A fourth screen may seem hard to accommodate, unless we account for our smart watches, our partners' devices, or whatever is left of the real world. We may indeed be ready for Neuralink, Elon Musk's brain-implanted AI, if only for the sake of simplifying and consolidating it all inside our minds, or permanent virtual reality to transport us to a different, less distracting reality. The metaverse is closer than we think. Whether we realize it or not, we are all looking at the world through the lens of AI now, while somehow managing to pay less and less attention to the real world.

If living in a state of permanent misfocus is the price we pay for indulging in the never-ending array of digital distractions, what do we actually get out of it? For any technology or innovation to become ubiquitous, it has to cater to our deep psychological cravings, as well as some evolutionary needs. There's something more profound going on here than being bored or feeling FOMO if we don't keep up with our digital updates. That something is probably our thirst for meaning, that is, our desire to make sense of the world and translate an otherwise chaotic, ambivalent, and unpredictable reality into a meaningful, familiar, and predictable world.

While we have always pursued the quest for meaning with indefatigable passion, the sources of meaning have tended to change quite a bit throughout the course of our evolutionary history. At first, we relied on shamans and the wisdom of the old, sage members of our group. We asked difficult questions and believed their answers. Then came mysticism and religion, followed swiftly by philosophy, science, and whatever cult or knowledge school you can think of. After philosophy, we got meaning from science, and there's always been art, music, sports, entertainment, and of course, experience. Life provides us with the raw ingredients to construct a vast catalog of meaning—we adore the conviction of having it and despair when we fail to find it.[38]

However, the AI age has complicated our relationship with meaning. On the one hand, it has given us too much of it, inundating us with sounds, images, and information that are specially tailored to our wants and needs. The wide range of stimuli we encounter every day is unprecedented in human history—from Facetuned images on Instagram to videos of car crashes, terrorist attacks, and insurrections on YouTube, and everything in between. More and more this barrage of stimuli is powered by algorithms that persuade us to click and like and buy, which implies validation of meaning or at least our craving for more.

On the other hand, we've become anesthetized to it all. As Gregory Robson at Iowa State University points out, the consequence of sensory overstimulation is often intellectual understimulation.[39] We may scroll through Instagram, browsing curated pictures of our friends and celebrities, but how often are we extracting real meaning from these experiences? In the early phase of the digital age, we were mostly concerned about the rise of distractions; now we have shifted to a phase in which life itself seems like an entire distraction. Our highest hope seems to be to become *someone else's* digital distraction, which, besides the self-congratulatory digital popularity metrics, does not make our own lives particularly fulfilling.

Historically, we have always depended on experience to shape our attitudes and values. We felt, thought, and formed our own opinions through impressions and lived experiences.[40] But when our experiences are limited to predigested algorithmic information designed to cater to our existing values or beliefs, we are deprived of the intellectual equivalent of metabolic processing, just as junk food consumption or processed food slows down our metabolism.[41]

We seem to find meaning only when we find a way to remove ourselves from the surplus of information, the ubiquity of symbolic overexposure, or the tedious repetition of everyday human experience. Finding more moments to be by ourselves, far from the digital crowds in the hopes that we can abstract our minds into deep thinking, may indeed be our best shot at reclaiming some of the meaning our lives seem to have lost. The choice is ours: to ignore the irrelevant rather than make it relevant by attending to it. Our tickets to the metaverse have not been booked yet, but we are neglecting life as we once knew it, to our own peril.

However, as the next chapter will show, the AI age has also disrupted one of the major mental mechanisms to resist attentional distractions, namely, our sense of autonomy and self-control. The same algorithms that are responsible for toying with our focus and concentration, tangling a never-ending array of digital carrots and virtual sweets, are eroding our patience and our ability to postpone gratification and make short-term mental sacrifices to enjoy long-term emotional and intellectual benefits.

TEST YOURSELF

Are You Unfocused?

Agree or disagree with the following statements:

- Your screen time continues to go up, in part because you keep monitoring your screen time stats.
- You rarely have lunch or dinner with someone without checking your phone.
- You wake up in the middle of the night to check your phone.
- You enjoy virtual meetings so long as you can multitask.
- During in-person meetings, you visit the restroom in order to secretly check your messages.
- You tend to connect to Wi-Fi on flights, even if you don't have work to do.
- You find it hard to concentrate or focus on anything for more than five minutes in a row.
- You have tried productivity apps, like distraction blockers, with little success.
- Your friends, partner, or work colleagues have complained about your smartphone addiction.
- You got distracted by other apps, sites, emails, or alerts while trying to complete this short assessment.

Add one point for every statement you agree with; then add up your points.

0–3: When it comes to distractibility, you are probably a cultural outlier, like someone still living in the 1980s.

4–6: You are within a range that can easily be flexed into the direction of more focus, attention, and less distractibility, though you will need to work at it.

7–10: You are the perfect customer of social media platforms and a cultural emblem of the AI age. Detoxing from your AI-fueled platforms and tools may be the only way to remind yourself of what you are missing while your attention is hijacked by digital technologies: that thing called *life*.

Chapter 3

The End of Patience

*How the AI age is making
us more impulsive*

The civilized man is distinguished from the savage mainly by
prudence, or, to use a slightly wider term, forethought. He is
willing to endure present pains for the sake of future pleasures,
even if the future pleasures are rather distant.

—Bertrand Russell

"Good things come to those who wait." Or do they?

Throughout human history, the most brilliant minds have high-lighted the power of patience. For example, Aristotle wrote that "patience is bitter, but its fruit is sweet." Tolstoy observed that "the strongest of all warriors are time and patience." Molière remarked that "trees that are slow to grow bear the best fruit," and Newton wrote that "genius is patience." But, in a world that is structured around the one-click-away premise, that famous aphorism—good things come to those who wait—sounds more like a pretext from people who miss deadlines or an excuse from the manager of a poorly run restaurant.

We no longer see patience as a virtue. Of course, we may admire those who exhibit it, but only because we are jealous of them. They're outliers—freaks of nature. It is similar to the way we hold humble leaders in high regard because the vast majority of leaders are utterly devoid of humility.[1]

In the AI age, we don't just lament our inability to focus and switch off but also be patient. Our impulsivity is out of control. Just a few seconds of buffering are enough to make our blood boil, and a slow connection is perhaps the best modern equivalent to medieval torture. For a species once celebrated for its self-control, agency, deliberation, and capacity to postpone gratification, we have digressed to the patience levels of an average five-year-old. No matter what we do or where we go, we can't seem to go fast enough.

Life in the Fast Lane

Our interest in going faster has fueled much of the technological innovation of the past two decades. Before AI, we said the same for fast food, the car, and pretty much every sphere of human consumption. Humanity, it seems, appears to prefer perpetual acceleration, living our lives in fast-forward mode.

Though how impatient or impulsive we are to begin with will predict our propensity to be addicted to WhatsApp, Instagram, and Facebook, our dependence on these and other AI-driven platforms will further increase our impulsivity levels, which explains why merely sixteen seconds of buffering are sufficient to trigger frustration in a typical modern adult.[2] (I don't know about you, but

I'm more of a three-second kind of person.) Recent academic studies report that a third of smartphone users admit to grabbing their phones within five minutes of waking up, which is probably before they reach out to their partner, and over 40 percent of us reportedly check our smartphones throughout the night.[3]

Since the dawn of the internet, addictive technology usage has been proposed as a new, stand-alone form of psychiatric pathology, and there's wide consensus about the fact that compulsive digital usage mirrors other addictions, such as gambling, drinking, or smoking, in both symptomatology and the range of psychological causes and consequences.[4] However, what was once considered pathological and unusual—for example, spending lots of time playing video games, shopping online, or being glued to our social media platforms—has simply become the new normal. Naturally, the internet needed to happen for internet addictions to exist, but in merely two decades our digital compulsions have managed to both emerge and be downgraded in the mental health risk scale from symptoms of weird and niche pathology to a mere cultural or social symptom of our times.

How far we've come. In the late nineties, we were all excited with the adrenalizing sound of those twenty-second intervals in which our computer attempted to establish a connection to the dial-up internet—the noise of being transported into a new world of seemingly endless possibilities and interactions with distant friends and relatives. Fast-forward two decades and a five-second delay in an incoming email, app refresh, software update, or Netflix buffering seems sufficient to ignite the desire to smash our screens or throw our devices through the window. While it's too soon to say whether AI is reducing our patience or killing it altogether, there's little doubt that major AI-related tools, innovations, and

gizmos are largely focused on speeding things up. AI is broadly a time-reduction tool, making things speedier and more efficient.

Before the AI age, we used to go out to bars and talk to people, have a drink, and evaluate whether we should exchange numbers, talk more, and establish whether we should meet up to get to know each other. Then came Tinder and Bumble, which diluted much of the creative burden hitherto assigned to romantic in-person encounters. Before the current pandemic, online and mobile dating had already emerged as the number one way to meet someone (39 percent for heterosexual Americans and much higher for LGBTQ dates).[5] A significant number of single—and not so single—people who visited bars did so having previously short-listed their matches through mobile dating apps, prompting some bars to offer "Tinder Tuesday" nights where 100 percent of guests had a virtual date lined up. Tinder claims that since its launch a decade ago, it has produced 55 billion matches, probably more than all spontaneous analogue matches in the history of modern bars.[6]

Think about the process of finding a job. We used to primarily rely on friends, colleagues, graduate recruitment fairs, or word of mouth and send job applications with the hope of an interview, an internship, or participation in a formal assessment center. Then came Indeed, LinkedIn, and HireVue, and the prospect of on-demand jobs that can be sourced almost like a Tinder date seems just around the corner. Gig-economy platforms like Upwork, Fiverr, and Uber still account for a tiny share of the total number of jobs in any economy, but things could change.[7] AI is vastly improving the experience of those services, too, using our past behavior, and the behavior of others that look or act like us, to offer an ever-expanding range of personalized choices that reduce the time

and friction we would otherwise spend on meeting new people or finding new jobs.

But the ability to quickly find a date or job does not increase our satisfaction with our choices, let alone our long-term success with them. You will need patience to give the date or job a proper shot, assess whether your choices made sense, and avoid FOMO while you are bombarded with other real or imaginary alternatives. It is no different from Netflix algorithms that instantly recommend what we should watch, only to have us think of all the other potential options we could be watching, according to its own recommendation engine. Even when life offers us an endless stream of possibilities and opportunities, including unlimited content streaming, we never have enough time to try it all, leaving us unfulfilled and ungratified because of our awareness of what we are unable to experience and what we are missing, at least in our minds.

This compulsive FOMO is most clearly highlighted in online shopping, with as many as 16 percent of American adults reportedly engaging in compulsive consumption patterns regularly.[8] Most sites use AI to curate, nudge, and upsell products to customers, creating sticky relationships between brands and consumers. The behavioral indicators that have been identified as key markers of compulsive online shopping look uncomfortably normal, suggesting that the actual figure may be significantly higher than 16 percent: "I spend a lot of time thinking or planning online shopping." "Thoughts about online shopping/buying keep popping in my mind." "Sometimes I shop online in order to feel better." "I have been bothered with poor conscience because of my online shopping/buying."[9]

These statements seem as ordinary and normative as "I often shop online," which does not in any way imply that online shopping is unproblematic. On the contrary, it indicates the pervasive impact that our online retail compulsions have had. I would not be surprised to see future regulations to protect consumers from these habits. The mayor of Belgium's third biggest city, Charleroi, recently called for abolishing online shopping in his country, describing e-commerce as a "social and ecological degradation."[10]

Not just government regulations but also cultural antidotes have emerged to contain or resist the fast and furious approach to life instilled by the AI age—from the slow food movement in northern Italy to the proliferation of three-hour movies or ten-season series (as if content duration were positively related to award nominations), and even long-term financial investment philosophies, such as Warren Buffett's "only buy something that you'd be perfectly happy to hold if the market shut down for ten years."[11] That's sound investment advice, but for some, buying GameStop stock on Robinhood is more fun than dumping cash into an index fund.

However, all these attempts at slowing down are dwarfed by the typical challenges imposed by modern life, notably our need to immediately access information, services, or people just by reaching for our phones. Consider how TikTok, which went from 100 million to 1 billion users in the last three years, with some countries reporting average usage time of three hours per day, uses AI to hook consumers, earning TikTok the title "digital crack cocaine."[12] Without requiring users to even report much about themselves, the platform can quickly personalize content recommendations and progressively leverage users' viewing patterns to increase content relevance. As analysts have noted, "TikTok is the first mainstream consumer app where artificial intelligence IS the product. It's representative of a broader shift."[13]

TikTok is highly effective, but it's not perfect. Like with any AI product, the quality of its algorithms depends on the army of users who train and refine it. When investigators at the *Wall Street Journal* created a hundred automated accounts, they discovered that, over time, "some of the accounts ended up lost in rabbit holes of similar content, including one that just watched videos about depression. Others were served videos that encouraged eating disorders, sexualized minors and discussed suicide."[14] If this minimal application of AI to increase the addictive nature of social media can be so impactful, it is scary to think what a fully fleshed and perfected AI engine could achieve.

Our Brain on Speed

If the AI age requires our brains to be always alert to minor changes and react quickly, optimizing for speed rather than accuracy and functioning on what behavioral economists have labeled System 1 mode (impulsive, intuitive, automatic, and unconscious decision-making), then it shouldn't surprise us that we are turning into a less patient version of ourselves.[15]

Of course, sometimes it's optimal to react quickly or trust our guts. The real problem comes when fast mindlessness is our primary mode of decision-making. It causes us to make mistakes and impairs our ability to detect mistakes.[16] More often than not, speedy decisions are borne out of ignorance.

Intuition can be great, but it ought to be hard-earned. Experts, for example, are able to think on their feet because they've invested thousands of hours in learning and practice: their intuition has become data-driven. Only then are they able to act quickly in accordance with their internalized expertise and evidence-based experience. Alas, most people are not experts, though they often think they are. Most of us, especially when we interact with others on Twitter, act with expert-like speed, assertiveness, and conviction, offering a wide range of opinions on epidemiology and global crises, without the substance of knowledge that underpins it. And thanks to AI, which ensures that our messages are delivered to an audience more prone to believing it, our delusions of expertise can be reinforced by our personal filter bubble. We have an interesting tendency to find people more open-minded, rational, and sensible when they think just like us.

Our digital impulsivity and general impatience impair our ability to grow intellectually, develop expertise, and acquire knowledge. Consider the little perseverance and meticulousness with which we consume actual information. And I say *consume* rather than inspect, analyze, or vet. One academic study estimated that the top-10-percent digital rumors (many of them fake news) account for up to 36 percent of retweets, and that this effect is best explained in terms of the so-called echo chamber, whereby retweets are based on clickbait that matches the retweeter's views, beliefs, and ideology, to the point that any discrepancy between those beliefs and the actual content of the underlying article may go unnoticed.[17] Patience would mean spending time determining whether something is real or fake news, or whether there are any serious reasons to believe in someone's point of view, especially when we agree with it. It's not the absence of fact-checking algorithms during presidential debates that deters us from voting for incompetent or dishonest politicians, but rather our intuition. Two factors mainly predict whether someone will win a presidential candidacy in the United States—the candidate's height and whether we would want to have a beer with them.

While AI-based internet platforms are a relatively recent type of technology, their impact on human behavior is consistent with previous evidence about the impact of other forms of mass media, such as TV or video games, which show a tendency to fuel ADHD-like symptoms, like impulsivity, attention deficits, and restless hyperactivity.[18] As the world increases in complexity and access to knowledge widens, we avoid slowing down to pause, think, and reflect, behaving like mindless automatons instead. Research indicates that faster information gathering online, for example, through instant Googling of pressing questions, impairs long-term knowledge acquisition as well as the ability to recall where our facts and information came from.[19]

Unfortunately, it's not so easy to fight against our impulsive behavior or keep our impatience in check. The brain is a highly malleable organ, with an ability to become intertwined with the objects and tools it utilizes. Some of these adaptations may seem pathological in certain contexts or cultures, but they are essential survival tools in others: restless impatience and fast-paced impulsivity are no exception.

Although we have the power to shape our habits and default patterns of behaviors to adjust to our habitat, if pace rather than patience is rewarded, then our impulsivity will be rewarded more than our patience. And if any adaptation is *overly* rewarded, it becomes a commoditized and overused strength, making us more rigid, less flexible, and a slave to our own habits, as well as less capable of displaying the reverse type of behavior.[20] The downside of our adaptive nature is that we quickly become an exaggerated version of ourselves: we mold ourselves into the very objects of our experience, amplifying the patterns that ensure fit. When that's the case, then our behaviors become harder to move or change.[21]

When I first returned to my hometown in Argentina after having spent a full year in London, my childhood friends wondered why my pace was so unnecessarily accelerated—"Why are you in such a hurry?" Fifteen years later, I experienced the same disconnect in speed when returning to London from New York City, where the pace is significantly faster. Yet most New Yorkers seem slow by the relative standards of Hong Kong, a place where the button to close the elevator doors (two inward-looking arrows facing each other) is usually worn out, and the automatic doors of the taxis open and close while the taxis are still moving. Snooze, and you truly lose.

There may be limited advantages to boosting our patience when the world moves faster and faster. The right level of patience is always that which aligns with environmental demands and best suits the problems you need to solve. Patience is not always a virtue. If you are waiting longer than you should, then you are wasting your time. When patience breeds complacency or a false sense of optimism, or when it nurtures inaction and passivity, then it may not be the most desirable state of mind and more of a character liability than a mental muscle.[22] In a similar vein, it is easy to think of real-life problems that arise from having too much patience or, if you prefer, would benefit from a bit of impatience: for example, asking for a promotion is usually a quicker way of getting it than patiently waiting for one; refraining from giving someone (e.g., a date, colleague, client, or past employer) a second chance can help you avoid predictable disappointments; and waiting patiently for an important email that never arrives can harm your ability to make better, alternative choices. In short, a strategic sense of urgency—which is the reverse of patience—can be rather advantageous.

There are also many moments when patience, and its deeper psychological enabler of self-control, may be an indispensable adaptation. If the AI age seems disinterested in our capacity to wait and delay gratification, and patience becomes somewhat of a lost virtue, we risk becoming a narrower and shallower version of ourselves.

Is There Hope?

Psychological science tells us that self-control, defined as the "mental capacity of an individual to alter, modify, change or override their impulses, desires, and habitual responses" is like a core mental muscle.[23] Although each of us is born with a certain predisposition—our foundational strength or potential—the more we exercise it, the stronger it gets. This means that we all have the capacity to develop higher levels of self-control to resist our digital temptations in order to finesse our focus and cultivate our patience.

Alas, there is also a catch: as decades of research by Roy Baumeister, the leading scholar in this area, show, self-control also *tires* like a muscle.[24] That is, the more effort and willpower needed to strengthen our self-control and resist temptations, the less energy reserves we have left. For example, if you spend all day thinking that you mustn't eat that cookie, the cookie will monopolize your willpower, depleting any energy to exercise self-control with regard to anything else. You have a limited amount of motivational energy to fuel your self-control: the more you use for one thing, the less you have left for everything else. The same goes for following diets, adopting a healthier lifestyle, or trying to be a better person. From the perspective of self-control, monogamy is a bit like vegetarianism: it may feel morally right and is surely a noble goal but does not quite eliminate the tasty aroma of charcoaled beef. And the same goes for resisting our technological temptations.

Clearly, the only way to overcome our mindlessly impulsive dependence on technology is to use less of it and replace some of our online time with offline activities. While the range of activities that we can pursue offline has steadily declined, there is one universal activity that has enormous benefits and is strangely underrated: sleep.

A healthy sleep routine, including good quantity and quality of sleep (e.g., a good balance of restful, deep REM sleep), can expand your energy reserves, freshen your mind, and improve your mental and physical well-being. In a recent meta-analytic study, within- and between-individual variability in sleep quality and quantity was positively related to self-control.[25]

Interestingly, AI is advancing the science of sleep, helping to detect sleep problems and improve interventions and treatments. Yet, our phones and their blue light are disrupting our sleep. But, hey, at least we have apps that track our sleep patterns.

Another great and widely available option to boost self-control is exercise, with research showing that asking people to engage in two weeks of regular physical activity or disciplined fitness reduces their impulse shopping habits.[26] Inevitably, you will need some self-control or willpower to engage in exercise to begin with, just like attempting to eliminate any bad habits will require a firm decision to get started, followed by some degree of commitment. However, such efforts will tend to pay back as the relationship between self-control and exercise is bidirectional: they both boost each other, so the more you exercise, the more you expand your mental stamina. Thus, scientific reviews show that long-term participation in exercise, as well as improved physical fitness in general, significantly enhance self-control.[27]

The point is that the ability of machines to control and manipulate us says more about our lack of willpower and weak motivational strength than the sophistication of AI. Likewise, just because AI is advancing its ability to perform logical computations at scale, expanding the range of real-world problems it can tackle doesn't mean we have to downgrade our own intellectual performance in everyday life.

However, as the next chapter highlights, a critical yet rarely discussed characteristic of the AI age is the pervasive irrationality and bias underpinning human thought, which is by far a bigger threat to the world than the advancement of machine capabilities. We may have wanted artificial intelligence but have encountered human stupidity instead.

TEST YOURSELF

Are You Patient?

Agree or disagree with the following statements:

- Few things are worse than a slow internet connection.
- Seeing the buffering sign for more than thirty seconds is enough to drive me insane.
- My days go by faster and faster, or at least it feels that way.
- I have been called impatient at times.
- Slow people irritate me.
- I have responded to emails faster than I should have.
- It takes me a great deal of effort to be offline.
- If I weren't always connected, I would have more time to engage in healthy activities.
- The digital age has made me less patient.
- I often feel like I struggle with self-control issues.

Add one point for every statement you agree with; then add up your points.

0–3: You are a fine example of someone who, despite all digital temptations and distractions, has managed to remain composed, serene, and in control of things. Congrats! You're a unicorn.

4–6: You're average. You can work on some things.

7–10: The algorithms love you. They rule your life.

Chapter 4

Taming Bias

How the AI age exacerbates our ignorance,
prejudices, and irrationality

It is useless to attempt to reason a man out of a thing
he was never reasoned into.

—Jonathan Swift

Humans are generally known for their rationality, logical thinking, intelligent reasoning, and decision-making—at least according to humans. These qualities have contributed to undeniable progress in the fields of science, as well as advances in engineering, medicine, and even AI. But let's be honest. Humans are also dumb, irrational, and biased. This is especially true when we're trying to win arguments, impress others, make rapid and impulsive decisions, and feel good about the quality of our decisions (and more generally, ourselves).

Decades of research in behavioral economics show that humans are endowed with a remarkable range of reasoning biases to help them navigate the sea of complexity and streamline or fast-track their interactions with the world and others without overstretching their mental capacity.[1] In the words of renowned neuroscientist Lisa Feldman Barrett, "The brain is not for thinking."[2] Our brains evolved to make quick predictions about the world in order to enhance our adaptations, while economizing and preserving as much energy as possible. And the more complex things get, the more we try to do this and simplify things.

The most common uses of AI do more to advance our ignorance than our knowledge, turning the world into a more foolish and prejudiced place. Think of all the ways social media plays to our confirmation biases. The algorithms know what we like and feed us news stories that tend to fit our established view of the world. As expansive as the internet is, with its diverse perspectives and voices, we all operate inside our own filter bubbles. For all the time we devote to determining whether AI is actually "intelligent," there appears to be one unquestionable feature about it, namely, that its focus is largely to increase humans' self-esteem rather than intelligence. AI algorithms function as a sort of motivational speaker or life coach, a confidence-boosting agent designed to make us feel good about ourselves, including our own ignorance, which they help us ignore. When AI algorithms target us with the stories we want to hear (and believe), they boost our confidence without boosting our competence. To paraphrase comedian Patton Oswalt, in the sixties we put people on the moon with computers less powerful than a calculator. Today everyone has a supercomputer in their pocket and they're not sure if the world is flat or if vaccines are filled with wizard poison.

The modern history of human intelligence is very much an apologetic journey of self-humiliation. We started with the premise that people are not just rational but ruthlessly pragmatic and utilitarian, so that they can always be expected to maximize the utility and rewards in their decisions and logically weigh the pros and cons, ultimately picking what's best for them. This was the phase of *Homo economicus*, or rational man. Humans were seen as objective and efficient creatures of logic who would always act in intelligent ways. But the behavioral economics movement shattered this myth, presenting a long list of exceptions to this rule, making bias the norm and objectivity the exception, if not a utopia. Sure, we may be capable of acting rationally, but most of the time we will act on instinct and let our biases drive our decisions, as shown by the innumerable list of mental shortcuts or heuristics that make irrationality a far more probable outcome than rationality.

Instead of following the logic of an argument or the trail of evidence, we simply direct the argument or evidence toward a preferred outcome. Most of the time people act in irrational ways, even if they are still predictable.[3] We act not as the impartial prosecutor of an investigation, but as the criminal lawyer of the guilty defendant, which in this case is our ego. While this view still represents today's consensus about human intelligence, things are more nuanced than behavioral economists suggested. If you look at modern personality psychology, it is clear that people are predictably irrational, but you still need to decode each person's unique patterns of irrationality to predict and understand their behavior. In other words, human stupidity—more than human intelligence—comes in many different shapes and forms, which we can ascribe to personality: the unique tendencies and biases that make you you. The only universal bias is to assume that we are less biased than other people.

Can AI acquire a personality? If what is implied by this is a certain style of biased decisions, or a recurrent but unique pattern of preferred adaptations to specific situations, then the answer is a definitive yes. For example, we could imagine a neurotic chatbot, prone to making pessimistic, self-critical, and insecure interpretations of reality, craving excessive validation from others, and disregarding positive feedback because things cannot possibly be as good as they seem. This chatbot—call it neurotic.ai—would have mastered the art of impostor syndrome and continue to prepare for college and work assignments more than needed, while remaining dissatisfied and hypercritical about its own accomplishments. Or a machine learning algorithm with high impulsivity, prone to making overconfident interpretations of data, and drawing bold and wild insights from very limited data points, insufficient facts, and so on. Perhaps this overconfident AI would end up being rewarded for these careless and overly optimistic inferences, very much like overconfident and narcissistic executives are celebrated for their arrogance and for being unjustifiably pleased with themselves, which reinforces these delusional biases. People are inclined to follow and respect others when they see them as smart, which is unfortunately influenced by a wide range of factors other than their actual intelligence. Delusional confidence is high up the list. In that sense, if AI succeeds at emulating humans, we may end up mistaking algorithmic overconfidence for competence.

We could also picture some form of unfriendly or selfish AI that compensates for its lower self-concept by bringing other people down and making negative evaluations of others, even if that means a poorer understanding of reality. This could include truly racist or sexist chatbots who take personal pride in making derogatory remarks about certain demographic groups so that they feel better about their own, though this would probably require assigning a gender, race, or nationality to the chatbot's identity. And perhaps we could even design an overly curious, inventive, and creative AI that makes unusual associations and focuses more on style than substance, imitating the poetic thinking tendencies of artists rather than deploying rigid mathematical thinking, and so on.

You're More Biased Than You Think

Most of us think of ourselves as less biased than we are and, of course, less biased than others. Liberals think that conservatives are the unwitting victims of mis- and disinformation. Conservatives think liberals are a threat to free speech. Most of us think that we surely don't hold racial biases, while others do. Or we think we see the world as it really is, while others see it with rose-tinted glasses. This is false.[4] If you genuinely disagree, you're probably just bullshitting yourself. If we were to ask a hundred people if they are biased, probably less than 10 percent of respondents would agree. However, if we asked the same hundred people if other people are biased, 90 percent of them would say yes.

We may *think* our intellect is guiding us to act in logical or rational ways, but our will is in charge. In the famous words of the German philosopher Arthur Schopenhauer: "Will is the strong blind man who carries on his shoulders the lame man who can see." Fittingly, Schopenhauer also wrote, "The world is my idea," a statement that epitomizes the rise of "subjectivity" as a core philosophical principle and prompted his usually serious academic colleagues to ponder, "What does his wife have to say about this?"[5]

One of the oldest findings in social psychology is that people interpret successful events as personal wins but blame unsuccessful events on external, uncontrollable circumstances, such as luck or cosmic injustice.[6] It's always your skill or talent when it works out well, or at least hard work and dedication, but bad karma, unfairness, or random freaks of nature when it doesn't.

Indeed, research shows that the vast majority of us indulge in what is known as the *optimism bias*. As my UCL colleague Tali Sharot states, "When it comes to predicting what will happen to us tomorrow, next week, or fifty years from now, we overestimate the likelihood of positive events, and underestimate the likelihood of negative events. For example, we underrate our chances of getting divorced, being in a car accident, or suffering from cancer."[7]

Yet, most of us don't see ourselves this way. Which isn't surprising since humans have a unique capacity for self-deception, and there's little one can do to persuade us otherwise.

We have a big awareness gap: most of us are willfully ignorant of our own biases, prejudices, and blind spots. But there's more bad news: even if we make ourselves aware of our own limitations, we may not be able to fix the problem. Look no further than modern interventions to de-bias the workplace, notably unconscious bias training, which is all the rage in HR circles. As a recent meta-analysis of nearly five hundred studies shows, with a great deal of effort, it is possible to produce very small changes in implicit or unconscious measures of attitudes and biases, but these changes have no meaningful impact on behavior.[8] Obviously, those who develop these programs mean well, as do the teams within organizations that host them. But they just don't work.

First off, they preach to the choir, or they appeal to those who think of themselves as "open-minded" or "liberal" or "not prejudiced." But there's a small chance these descriptions are true; usually, they're simply naive and overconfident strategies in self-deception. They push us to blame others. And when we point the finger at others for being biased, we are implying that it's *they* who are the problem rather than us or the system.

If you want to *control* your behavior, it may be beneficial to understand not only your attitudes but also how others may judge them, and what behaviors they will inspect to infer your beliefs. But the problem is these de-bias programs naively assume that the awareness of our own biases will lead us to act in more open-minded ways. If only. And, by thinking this way, we're failing to promote accountability and fairness while inhibiting our ability to evolve. This is also why family dinners can be so frustrating. It doesn't matter how crazy your dad is. No degree of evidence or proof will challenge his deep values or core beliefs. Plus, he's sure to counter your facts with his own facts. This is the defining irony of our times. The more data and information we have access to, the easier it is to misinterpret or cherry-pick data that confirms your beliefs. This is how different countries or leaders made totally different interpretations of pandemic data, from "this is like a flu," to inaccurate predictions about its effect on the economy, the housing market, and mental well-being.[9]

If we truly want to become more rational, more inclusive, and less biased, we should worry less about what we really think or believe in, and be more open to accepting or at least trying to understand what others believe. There's little use in trying to police people's thoughts or ideas. Instead, we should try to behave in loving or at least polite ways. If hate is not directed at others, it is less likely to be directed inward. Research indicates that even if we have to force ourselves to behave in prosocial or kind ways to others, this will positively impact our mood and self-concept, making us more open-minded in turn, and that random acts of kindness have the ability to boost our empathy and altruism.[10]

Kindness forces us to reconceptualize our self-views and reframe our self-concept, holding ourselves to higher moral grounds and ideals. So, even when we donate to charity with the primary purpose of *seeming* generous to others, we will end up seeing ourselves as good people, which in turn will promote good behaviors going forward. Although most digital echo-systems, particularly social media, foster impulsive reactions to people's actions, which result in a great deal of hostility, trolling, and bullying, it is generally easier to display kindness and consideration online than offline. For starters, the opportunities to pause, reflect, and exercise self-control are much higher than in in-person interaction. And the incentive is higher: anything you do online will be recorded and registered in perpetuity, and the entire world (well, at least your world) is watching. So, all you need to do is to stop yourself from reacting or responding until you have something positive to say—I know, easier said than done.

Can AI Help Us?

Interestingly, AI may be able to help us on the bias front. One of its biggest potential utilities is to reduce human biases in decision-making, which is something modern society appears to be genuinely interested in doing.

AI has been successfully trained to do what humans generally struggle to do, namely, to adopt and argue from different perspectives, including taking a self-contrarian view or examining counterarguments in legal cases.[11] In general, you can think of AI as a pattern-detection mechanism, a tool that identifies connections between causes and effects, inputs and outputs. Furthermore, unlike human intelligence, AI has no skin in the game: it is by definition neutral, unprejudiced, and objective. This makes it a powerful weapon for *exposing* biases, a key advantage of AI that is rarely discussed. Here are a few examples.

Example 1: An online dating site with millions of users who report their romantic (and sexual) preferences by constantly training the algorithms to predict their preferences uses AI to discover what most men (straight or not) and women (straight or not) generally "optimize for." In the process, AI improves the recommendations so that users have to devote less time to picking a potential date or partner.

Example 2: An online search engine that feeds people content (news, media, movies) deploys AI to detect what preferences viewers have by simply feeding people things that other people who are similar to them in either demographic or other preference features have. It quickly learns to effectively serve people the content they are most likely to consume and least likely to resist.

Example 3: A sought-after employer with millions of job applications per year leverages AI to compare the characteristics of job applicants with those of their employees, optimizing for a high degree of similarity in the profile of new job applicants and incumbents or current employees who have succeeded in the past. In essence, the more you resemble people who have historically done well in the company, the higher your likelihood of being selected for a job.

This is all great. But herein lies the problem. AI and machine systems are only as good as their inputs. And if the data we use as input is biased or dirty (we were all excited by big data until we realized this was mostly mean, dirty data), the outputs—the algorithm-based decisions—will be biased, too. Worse, in some scenarios, including data-intensive technical tasks, we trust AI over other humans. In some cases, this may be a valid response. But you can see the problem if a system is spitting out biased decisions and we blindly trust the results.

Yet, this problem also highlights the biggest potential AI has for de-biasing our world. But it does require an understanding—and willingness to acknowledge—that the bias is not the product of AI, but rather, only *exposed* by AI. To stay with my examples: in the first, if you don't use AI or algorithms to recommend to online dating users who they should date, their preferences may *still* be biased (e.g., people who belong to their own ethnicity, age, nationality, socioeconomic status, or attractiveness group, not to mention height). In the second example, the only alternative to "giving people what they want" (to read, hear, and see online) would be to give them what they *don't* want, which is an evolution in the moral consciousness of advertising and media targeting, but perhaps not best for the survival of for-profit corporations. In the third example, refraining from using AI to select and recruit candidates that fit a certain mold (say, middle-aged white male engineers) will not stop people who fit in with that tribe from succeeding in the future. If the bias does not go away just because you don't use AI, then you can see where the bias actually lies—in the real world, human society, or the system that can be exposed through the use of AI.

Consider these two scenarios: a racist manager could mitigate his bias by hiring people on the basis of their algorithmic score on a job interview, using AI to pick up relevant interview cues to predict the candidate's future job performance, all while *ignoring* their race, which humans find impossible to do. That's an ideal scenario (though in an ideal world, of course, we would not have a racist manager in the first place).

Now, imagine a nonracist manager who may rely on an algorithm to automate the prescreening of candidates based on their educational credentials, previous job experience, or likelihood of being promoted. That may sound optimal, but it could be highly problematic. For any AI system or algorithm to learn, it needs to ingest data sets including labels, such as "cancer" or "noncancer," "tree" or "traffic light," and "muffin" or "Chihuahua." But when these labels are the product of subjective human opinions, such as in the case of "good employee" versus "bad employee," it should not surprise us that AI will learn our biases. In this case, the use of the algorithm could end up having adverse impact on minority candidates, prompting the manager to inadvertently make racist selections and, to make matters worse, assume their decisions are objective.

This is, indeed, how AI has failed in the past—through contamination in the training data set or by making objective decisions on the basis of unfair, flawed, unethical historical decisions. So, when Microsoft tried to deploy a Twitter chatbot to engage millennials, (human) Twitter users quickly trained it to use foulmouthed language and post racist and sexist tweets—no prizes for guessing where the dark side was engendered, that is, human versus artificial intelligence.[12] The fact that humans get a kick out of making chatbots do antisocial, sexist, racist things clearly says little about AI's dark side and a great deal about the dark side of human psychology. If reading this paragraph prompted you to check the racist and sexist tweets of Microsoft chatbot Tay, the rule is rather valid for you, too. Likewise, when Amazon decided to scrap its recruitment AI on the basis that it recommended many more male than female candidates for its open job vacancies, it is clear that eliminating the AI would not automatically eliminate the disproportionate number of *male* programmers who succeed relative to *female* employees.[13]

Thus, most high-profile cases of AI horror stories, or attempts to transfer human decision-making to machines, are akin to "shooting the messenger." The very algorithms that are indispensable for exposing the bias of a system, organization, or society are lambasted for being biased, racist, or sexist, just because they do a terrific job replicating human preferences or decision-making. If only AI could convert people into a more open-minded version of themselves by showing them what they don't want to (but perhaps need to) hear, it would certainly do that. If I'm a neoconservative libertarian, AI could show me socialist or left-wing progressive content to increase my empathy for the left or switch my political orientation. If my music-listening habits reveal a very white, middle-aged range of preferences, and AI could expose me to young, hip, Black, urban music, it could systematically change my taste.

If AI alone would present to hiring managers people who are categorically different from those they have hired in the past and change the managers' preferences, then we would not talk about open-minded AI or ethical AI, but open-minded humans or ethical, intelligent, curious humans. It's the same for the reverse, which is the real world we live in.

The risks of AI breaking bad or algorithms going rogue can be mitigated if ethical humans remain in the loop but exacerbated when humans lack integrity or expertise. Much of the outcome is determined by *our* (human) understanding of what we are actually asking AI to do. If, again, we ask algorithms to replicate the status quo, and this combines a meritocratic or open-minded illusion with political, nepotistic, prejudiced, or biased forces, the main contribution of AI will be to refute the idea that what we have is fair, unbiased, and meritocratic.

If, conversely, ethical and competent humans are involved in the process of vetting, curating, and cleansing the training data that will fuel AI, then there's tremendous opportunity to use AI as a tool to diagnose and expose biases, and actually overcome them. Therein lies one of AI's big contradictions: what began as a tool to compete with human intelligence has the potential to reduce human bias but also risks exacerbating our flawed human nature by eroding our good bits and magnifying our bad bits.

At least humans are still very much in the driving seat, that is, in control of what AI is or is not used for. Opposition to AI, whether from the general public or actual AI users, is at its peak when AI recommends different decisions, behaviors, and choices than what humans intuitively prefer: for example, watch this movie, hire this person, go to this restaurant, buy these sneakers. When AI aligns with our preferences while exposing their dark side, we are quick to blame AI for our own inner demons, instead of acknowledging our own biases.

In this way, AI could become the biggest *reality check* weapon in the history of technology but is instead co-opted as a reality-distortion tool. To the degree that AI can help us confirm our own interpretations of reality or make us look good, we will embrace it. Failing that, we should regard AI as a failed experiment.

Reality Bites

We all believe what we want to believe. Why? Because our delusions are comforting. They help us replace an unpleasant version of reality with one that is soothing and compatible with our generous and lenient self-views.[14]

To combat our own self-delusions, we need to be less confident in our own views, opinions, and knowledge. Asking questions is more important than having answers. And as Stephen Hawking famously noted, "The biggest enemy of knowledge is not ignorance, but the illusion of knowledge."

We also need to be willing to accept feedback from others, which closes the gap between how we see ourselves and how others see us.[15] But that's a tall task, as the AI age has diluted feedback to a meaningless, repetitive, and semiautomatic ritual that produces positive feedback loops. So, for instance, when we post stuff on Facebook, Snapchat, TikTok, Twitter, or Instagram, it's not hard to get likes because liking something is a relatively low-energy, low-cost thing to do. Most people will like it, even if the feedback is fake, and it's likely to result in reciprocity later on. In the early days of LinkedIn, some people collected long endorsements from others, which they then reciprocated, so the endorsements said more about your friends than your skills or talents. However, this makes feedback far less useful than it should be. Facebook took over a decade to finally decide to include a "dislike" button, though Mark Zuckerberg described it as a function to express "empathy."[16] Relative to the positive feedback function, it is hardly used. Whatever we post, people will either like it or ignore it, but they will probably not dislike it. We may not be getting any real feedback from others, and instead we are being bombarded with fake positive feedback.

We're also incentivized to ignore the little critical or honest feedback we may actually receive. Think about the job or role where helping others with constructive feedback matters most, namely, leadership. Research shows that managers find it extremely hard to provide employees with negative feedback on their performance, which is why employees are often surprised when they fail to get a promotion or a bonus, and even when they are fired.[17] At the other end, managers and leaders are prewired to ignore negative feedback because they are not self-critical and prefer to surround themselves with people who suck up to them. The more you suck as a leader, the more you can expect people to suck up to you.

So, when we hear that we shouldn't worry about what people think of us and that if we think we are great, we probably are. We are the hero we are in our mind. The AI age has turned us all into a mini version of Kim and Kanye: we can all create echo chambers where even our most trivial and meaningless comments are celebrated and glorified, and where we are actually rewarded for behaving in a self-centered way, showcasing our egotistical self-obsessions, and for indulging in inappropriate self-disclosure. The best way to get fans and followers is to become your biggest fan. Not even our parents thought as highly of us, though they surely have their quota of responsibility for inflating our egos.

Academic research indicates that people who function effectively interpersonally and have accurate self-perceptions tend to incorporate other people's opinions into their sense of self, which runs counter to the idea that we should just be ourselves and ignore people's perceptions of us.[18] The ability to present ourselves in strategically and politically astute ways is indeed critical to succeeding in any professional context.[19] Those who live by the mantra "don't worry too much about what other people think of you" are rarely positively viewed by others. Academic reviews have highlighted that successful people worry a lot about their reputations, and they care deeply about portraying themselves in a socially desirable way.[20] When we enjoy the luxury of convening with colleagues and clients in person, we allow others to gain impressions of us based on our physical presence in a three-dimensional space, including our handshakes and our voices, which, as author Erica Dhawan explains in *Digital Body Language,* is largely replicated in virtual settings.[21]

The trouble is the alternative isn't much fun. Being your own harshest critic, judging your actions through a demanding and perfectionistic lens, is the exact opposite of what anyone would do if even remotely interested in enjoying life.

And even though self-awareness is a necessary driver of effective personal development, it is insufficient. It is perfectly possible for someone to become self-aware and gain a deeper level of self-understanding and yet not improve.

The Human Ethics of the AI Age

Much of the discussion around AI centers on the issue of ethics, with the most common assumption being that machines are going to either become evil and turn against us—presumably because they are amoral at best and immoral or worst—or replicate our worst character traits (of the two options, this is the best-case scenario). These fears are often based on the assumption that since humans created AI based on a human moral code, it will by definition misbehave or do wrong.[22] However, there's much more to human nature than the capacity for evil, and since AI can cherry-pick what elements of human behavior to emulate—mimicking some, but avoiding other traits—AI is at least capable of recreating benevolent or ethical human actions, or even perfecting them.

The question of AI morality is complicated by the fact that machines are incapable of acting ethically or unethically except according to human standards. In that sense, judging the moral character of AI is like judging the moral character of a dog: an anthropomorphic projection, unless our intention is to judge the moral character of the dog's owners. Humans project a sense of ethics onto their own and others' behaviors, including machines. So, when we look at decisions made by machines and we don't approve of them morally or ethically, because they fail to "align" with our own human values, we are mostly judging the humans who programmed the machines.[23]

Ethical questions always go back to humans, even if the behavior seems to be autonomously generated by machines or results from cascading errors inadvertently learned by machines that were emulating humans. The foundations of any ethical code are essentially human because we are always examining things from a human perspective. In that sense, horror stories about AI exterminating humans because we interfere with its objectives are not so much an example of unethical or evil, but single-minded AI. For example, say we program AI to produce as many paper clips as possible, and AI learned that in order to attain this goal, it must control a range of resources and components that would cause the extinction of humans, or even directly eliminate humans for interfering with its goal. This says less about AI's lack of empathy or moral sympathy for humans and more about its superpowers—assuming it could actually achieve this.[24] From an ethics standpoint, the closest human equivalent may be extinguishing certain animals because they are tasty or make good hunting trophies, or destroying the planet because we enjoy flying to in-person business meetings or cooling our offices with air-conditioning.

The stakes are higher when we are tasked with designing or programming machines that will replicate or reproduce human decisions at scale. Here again, though, AI is as ethical or unethical as your dog. You can reward your dog for doing certain things, like waiting patiently for her cookie, and punish her for others, like peeing in your living room. But there is no likely accusation of immorality when the dog violates any of the rules we taught her. If we train the dog to attack white people or women, then it is surely not the dog who's being unethical. If you would rather have a robot dog, such as Sony's Aibo, you can expect it to come with certain preprogramed behaviors straight out of the box, but in addition to these preprogrammed factory settings, it will also learn to adapt your personal ethics, thus incorporating your own moral standards. By the same token, any technology can be put to good or bad use, depending on the intention and ethics of the human and the ethical parameters we use to judge such intentions. If, in the paper-clip apocalypse problem, the tragic outcomes are caused by unforeseen programming consequences, then that says more about human stupidity than human morality, let alone artificial *intelligence*.

Consider the personal genomics and biotech firm 23andMe, which translates human saliva into genetic predispositions and medical profiles. While genetic profiling sounds rather frightening and off-putting to many people, not least because of its associations with Nazi eugenics or ethnic cleansing, the reality is that there are many possible applications for this technique, which will likely differ in their perceived morality. At the ethical or moral end of the spectrum, we can think of personalized medicine for the effective treatment of highly heritable illnesses. In particular, the technology 23andMe uses—single nucleotide polymorphism genotyping— could be deployed to help patients benefit from highly customized and targeted treatment for some of the 1 percent rare medical conditions that are about 99 percent genetic.[25]

Perhaps less ethical, but still not necessarily immoral, is the application of personal genomics to the auto insurance industry: customizing your policy on the basis of your character or personality. For example, traits such as conscientiousness, self-control, and reckless risk-taking are partly genetic, and they also predict individual differences in driving styles and performance. One of the reasons women end up in fewer road accidents and should be cheaper to insure than men is that they are generally more conscientiousness and less reckless from a personality standpoint. Objecting to any type of probabilistic segmentation or stochastic personalization simply means safer drivers will end up subsidizing their reckless counterparts, which is arguably unfair. Furthermore, since personality simply influences driving patterns rather than determining them, giving drivers feedback on their personality may help them adjust their behavior to correct bad habits and inhibit risky tendencies. Measuring how well people drive, especially when they are improving on their default tendencies, would end up overriding the earlier genetic predictions to increase not just fairness but also accuracy.

Even controversial applications of AI could be redirected toward slightly different goals to improve their ethical implications. For example, when a Facebook algorithm is trained to detect people who, because of their preferences data and personal behavioral patterns, may be regarded as probably undecided in their political orientation, the goal may be to inform or misinform or to nudge them to vote for X, Y, or Z or remain undecided altogether. That those potential voters are bombarded with fake news or that the additional "spin doctor" service of Cambridge Analytica included a range of non-AI-related methods, such as embezzlement, bribery, and the framing of politicians with prostitutes, does not make the company evidently ethical. Equally, if Facebook breached any data confidentiality or anonymity by selling or harvesting personal data to other parties, that is a matter that existing laws and regulations can address by typically targeting humans rather than algorithms. In short, the most algorithms can do, whether for Facebook, 23andMe, or Cambridge Analytica, is to identify patterns in the data: people who do or have X are more likely to do or have Y, without judging whether that makes them good or bad.

You can be sure that the AI tools used in digital political targeting are politically more neutral than Switzerland. Even if AI did play a role in determining the outcome for the 2016 Brexit referendum and the Trump election, the algorithms deployed did not truly care about Brexit or Trump, not least because they had no real understanding of this—perhaps like many of the humans who voted for or against these outcomes. Part of the scandalous reaction pertaining to the alleged digital or algorithmic interference with the 2016 EU referendum in Britain and the US presidential election is caused by critics' disdain of the result, because they consider Brexit and Trump a less moral or ethical outcome of those democratic elections. As a dual UK and US resident, I am myself part of that liberal group but accept the fact that my opinion may not be shared by around 50 percent of the voters in both elections.

Algorithms do pose a threat, as does any application of AI. If you use technology to standardize or automate unfair or unethical processes, you will simply augment, if not automate, inequality.[26] This happens when credit scores deny someone credit because they fail to meet certain criteria, or when insurance algorithms leave people unprotected because of a probabilistic error in their computation.

It is hard to be ethical without understanding the meaning of *ethical* to begin with, though of course one could be ethical by accident, especially when you don't understand immorality or un-ethics either. Discussions of ethical issues tend to default to legal matters, which are obviously pretty imperfect as a moral compass. Consider that homosexuality was listed as a psychiatric disorder in the United States until 1973, and that slavery was only abolished in America in 1865.[27] In the 1950s, it was illegal to sell your house to a Black person, and as Ronald Reagan noted, "Personally I would sell my house to anyone and you can buy it if you like, but it is only fair that people have the freedom to decide who they sell their house to." Is this ethical? By Reagan's standards, it is perhaps not totally unethical, let alone surprising. A simple yardstick to measure the ethical nature of our actions comes from Kant's famous categorical imperative: What would happen to the world if everyone acted like you? Would it be a better or worse place? Would it go up or down in the intergalactic ratings of Transparency International?

Ethics are a complex topic, and there's no shortcut to determining what is right and wrong without opening an arduous and potentially unsolvable debate, which would at best result in a cul-de-sac of religious, ideological, or cultural differences we call "moral conviction." But the only hope to improve fairness and well-being in our civilization is to agree on certain basic parameters that can provide the foundations for a moral, legal, and cultural code of action. Ethics are the governance framework that make one society more appealing and less toxic than others. We can measure it by the well-being of the poorest and most disenfranchised members. For example, a society in which it is virtually impossible for someone born poor to become rich must be questioned from an ethical standpoint, just as a society in which the rich abuse power and control systems via status or privilege must be held to ethical scrutiny. As Petra Costa notes, it is hard for democracy to function properly unless the rich feel somewhat threatened by the poor.[28]

TEST YOURSELF

Are You Biased?

How biased are you, especially compared with others?

- I am rarely unsure about things.
- I'm a black-or-white kind of person.
- I am at my best when I make quick decisions.
- I'm a highly intuitive person.
- Most of my friends have the same political orientation.
- I may have some biases, but I am less biased than the average person.
- I can read people like a book.
- I have never been impacted by fake news.
- My decision-making is always rational.
- It is important to work with people who share your values.

Add one point for every statement you agree with; then add up your points.

0–3: You are largely immune to the common biases and pervasive reality distortion that permeates our age (either that, or you have successfully deceived yourself into thinking that you are incredibly open-minded).

4–6: You may consider yourself average and within a range that can easily be flexed into the direction of more bias or less.

7–10: You are the perfect customer of social media platforms and a cultural emblem of the AI age, particularly if this high score surprises you.

Chapter 5

Digital Narcissism

*How the AI age makes us even
more self-centered and entitled
than we were already*

Legions of lusty men and bevies of girls desired him.

—Ovid

In the original mythological version that gave name to the trait, Narcissus, a handsome but emotionally detached and pompous young man, is punished by Aphrodite, the goddess of love, for his refusal to love anyone. His curse is to love only himself, so he ends up drowning while admiring his own reflection on a lake.[1]

The moral of the story? While a little self-love is to be expected, if you love yourself too much, you will have no interest in other people, which will harm your ability to function as a well-adjusted member of society.

Unsurprisingly, researchers have studied the behavioral consequences of digital technologies, including their relationship
to narcissism, a psychological trait associated with a grandiose
and inflated sense of self-importance and uniqueness, which tends
to reduce people's ability to tolerate criticism, care about others,
and accurately interpret reality, particularly their own abilities,
achievements, and failures.[2] Even if you are not part of the 2 percent
to 5 percent of the population that can be expected to meet the
medical criteria for pathological or clinical narcissism according
to psychiatric diagnosis, the AI age has normalized narcissism by
legitimizing the public displays of our egotistical and self-obsessed
nature. In that sense, we are all digital narcissists or are at least
nudged to behave like narcissists when we are online.[3]

The Rise of Narcissism

For over a century, prominent writers and social scientists have
warned that we're living in a self-obsessed era, a narcissistic epidemic, and that younger generations can only be described as the
me generation.[4] While it may be easy to dismiss these claims as
alarmist—and they probably were for past generations—there's evidence that narcissism is on the rise.

Psychologist Jean Twenge has tracked generational changes in
scientifically validated measures of clinical narcissism. For example,
one of the questions asked by these surveys is whether people think
they're destined to be famous. In the 1920s, only 20 percent of
the overall population answered yes. By the 1950s, the figure increased to 40 percent; in the 1980s, it went up to 50 percent; by the
early 2000s, it had risen to 80 percent. This suggests that, just as

by today's standards somebody who was deemed narcissistic in the 1950s would seem very modest and low-key, in 2050 we may look back and find even people like Elon Musk, Kim Kardashian, and Cristiano Ronaldo quite inhibited and private.

Narcissism, whether in its clinical or subclinical (i.e., lighter, adaptive, and much more pervasive) form can be understood as an extreme quest for self-enhancement, in the sense that everything narcissistic individuals do is motivated by a strong desire to inflate their self-views, lubricate their egos, and pamper their high opinion of themselves.[5] This includes the tendency to compare themselves with less successful people in order to validate their own self-concept, and the proclivity to evaluate their own talents (e.g., job performance, attractiveness, leadership potential, intelligence, etc.) in an unrealistically positive way, especially compared with how other people evaluate these.[6]

One of the key facets of narcissism is *grandiose exhibitionism*, which is characterized by self-absorption, vanity, and self-promotional impulses and is especially well-suited to a world in which human relations have been transferred almost entirely to digital environments. More than anyone else, narcissistic individuals feel the constant need to be the center of attention, even if the means to achieving this is to engage in inappropriate, awkward, or eccentric interpersonal behaviors. Back to Elon Musk, who wasn't content with monopolizing so much attention on Twitter, so he offered $44 billion to buy the entire business (then getting even more attention by pulling out of the agreed deal).[7]

Although most studies on digital narcissism report a correlation rather than causation, evidence points to a bidirectional link between narcissism and social media use. In other words, the more narcissistic you are, the more you use social media, which in turn makes you more narcissistic. Furthermore, experimental and longitudinal studies, which unlike correlational studies can detect causality, indicate that social media sites do inflate people's self-views.[8]

The AI age has given us a safety cushion. We can safely fish for compliments and seek praise without fearing rejection, even if it requires engaging in exaggerated self-promotion while actually being ashamed of our true self and pretending that others actually believe they are seeing the real version of ourselves.[9] The feedback we get from others reinforces the notion that our public persona is somehow real or genuine, which distances us further and further from who we genuinely are. Meanwhile, social media pays lip service to "authenticity," as if we were truly encouraged to act in natural or uninhibited ways instead of carefully curating our online persona. The American novelist Kurt Vonnegut once noted that "we are what we pretend to be, so we must be careful what we pretend to be." In the AI age, our digital persona has become the most emblematic version of our self, and its most generalizable feature is narcissism. If we are not narcissistic, we appear to pretend to be.

Of course, we cannot fully blame social media for making us narcissistic. After all, without a species already obsessed with itself, none of the technological platforms, systems, and innovations that fuel the AI age would exist in the first place. Had it not been for our constant self-focus, AI would be starved of data, and AI with no data is like music without sound, social media without the internet, or attention-seeking individuals without an audience.

Luckily for AI—and all of those who profit from it—there is no shortage of self-absorbing and egotistical activities to feed it: for example, posting selfies, sharing thoughts, engaging in inappropriate levels of public self-disclosure, and broadcasting our feelings, views, attitudes, and beliefs to the world as if we were the center of the universe or everyone else truly cared about them.[10] If the algorithms analyzing us were humans, they would surely wonder how a species so pathetically self-obsessed and insecure could have managed to get this far.

Consider a study that scanned people's brains to measure how they react to feedback on their selfies on social media. The researchers manipulated feedback—likes versus no likes—and tracked the level of psychological distress that subjects experienced while performing a challenging cognitive task. People exposed to positive feedback experienced less psychological distress, suggesting that social media approval can help narcissists alleviate the pain from social exclusion.[11]

Though we may think that the problem is with a small number of pathological users, we must point the finger at ourselves, too, and our self-aggrandizing needs as a human species. In this way, social media platforms are a lot like casinos. Even if you're not a compulsive gambler, if you spent enough time in a casino, you would probably place bets or play the slots. The same is true for social media. Though we may criticize those who show narcissistic tendencies on Snapchat, Facebook, and TikTok, these platforms encourage us to act in similar ways. Yet, just like spending more time at the casino won't improve our luck or winning streak, social media won't solidify or boost our self-concept. It does the opposite and tends to increase our insecurities.

Still, it will create an adrenalizing buzz or faux popularity and appreciation that will temporarily help us feel good about ourselves, exchanging our time and attention for ephemeral digital love bites. In this way, your pro-social instincts can be co-opted by algorithms that make it easy for you to accumulate "friends," churning out new connections like an Amazon recommendation engine suggests new sneakers, boosting your social status in the process, just like narcissistic individuals seek to widen or deepen their networks in order to satisfy their vanity levels and massage their egos rather than because of any real interest in people.[12] As such, the "social" aspect of social media resembles the antisocial aspect of the real world.

If people were forced to choose between a phone with a selfie camera or one with a traditional camera, you would have to bet on the former outselling the latter. The selfie is a dominant form of photography in the AI age, with global estimates suggesting that one person dies each week as a direct consequence of taking selfies (e.g., hit by a car, attacked by a thug, and falling from a rooftop).[13] But throughout the evolution of photography, self-portraits were the exception rather than the norm. Of course, famous visual artists (e.g., Velázquez, Rembrandt, Van Gogh, and Modigliani) did produce self-portraits, but they were generally more interested in those aspects of the world that did not include themselves. The thought that if they were alive today, they would spend most of their time posting selfies on social media does not bode well for anyone convinced that technological advancement equates to cultural progress or evolution.

The digital world encourages behavior that would never fly in the real world. In the real world, if you spend all your time talking about yourself and sharing everything you do and think with everyone else, with no filter or inhibition, people will leave the room and, unless you are their boss, give you hints that you are being obnoxious. But on Facebook or any social media platform, the worst that can happen is that people ignore you, without anyone else really noticing it. The much more likely scenario is that they will at least fake like what you do, reinforcing your shameless self-promotion and inappropriate self-disclosure with the help of algorithms that promote you for self-promoting.

The metrics of engagement, and the algorithms deployed to increase the time we spend on these platforms, are inherently egotistical—think of them as narcissistic nudges.[14] So, we are encouraged to share content, ideas, and media in order to get others' approval, like an insecure egomaniac who needs people's validation to maintain an inflated self-concept. We are permanently boasting, acting, and engaging in inappropriate self-disclosure in order to make an impact on others. A natural consequence of this is a recent psychological phenomenon defined as "broadcast intoxication," which takes place "when an individual experiences aspects of their self-esteem and social valuation by the reviews and reactions of others on social media."[15]

Although there are no objective societal advantages to loving ourselves, it is obviously rewarding to do so and certainly more pleasant than the alternatives—questioning or hating yourself. But from an evolutionary standpoint, a healthy self-esteem should function as an accurate indicator of one's social worth or reputation, signaling whether *others* accept, value, or appreciate us.[16] Our self-esteem evolved to tell us when and how we need to change our behavior so that we can do better in life. For example, if my ego is wounded because I get a low grade on a school exam, or I fail to get a job offer I desperately want, or my girlfriend breaks up with me, I am presented with great opportunities to repair these wounds to my ego by bouncing back from these setbacks, but it does require me to accept them in the first place. These opportunities are certainly less likely to emerge if my reaction is to be in denial about these failures.

Alas, when I'm starved of negative feedback, and living under the illusion that everything I do is admirable, courtesy of an avalanche of positive self-esteem-boosting signals from social media likes and other fake positive feedback, it is too easy to end up with a distorted self-concept and to become quite addicted to these self-enhancing psychological lubricants.

The Bright Side of Vanity, Vanished

Humans have always displayed a deep desire for appreciation, which is not only responsible for much of the vanity and entitlement in the world, but also civilization and progress. Our cultural evolution is fueled by the self-important motives and stubborn vanities of a small number of great individuals who are disproportionately responsible for driving the changes, innovations, and institutions that reshape and improve our world. The Medicis and Vanderbilts didn't have Twitter, but there's no reason to believe that their need for recognition and egos were any smaller than those of Elon Musk or Bill Gates. Human progress and innovation in any field is not just the story of great people, but also the material manifestation of their God complex.

Traditionally, however, the drive of exceptional achievers required not just a clear dose of unfulfilled vanity and self-importance but an even bigger dose of genius, brilliance, and grit, not to mention the ability to keep their egos in check in order to manage other people effectively so that they become a high-performing team. Whatever you think of cathedrals, symphonies, and hundred-year-old corporations, they are rarely the product of pure narcissism, but rather a watered-down version of it, diluted with hard work and competence, including leadership talent. It is this full combination of ingredients that has always represented the human algorithm underpinning extraordinary achievements, and that explains why even the God complex that may have driven the Rockefellers and Carnegies, and may drive Bezos and Musk today, still leaves society with a surplus of innovations that advance people's quality of life and changes the world for the better. You can look at this as a form of benevolent or altruistic narcissism.

Yet, excessive selfishness or greed is the parasite that corrodes our ability to function collectively as a well-oiled and cohesive social unit. Greed is a major cause of inequality, not because of its valuable accomplishments, but because of its tendency to crave an excess of power, status, and control, and, in turn, undermine democracy.[17] When left unchecked by laws or empathy, greed is the reason why humans self-destroy and destroy others, why institutions and entire states fail, and why inequality mushrooms in every period of history.

The world has never been wealthier than today, but it has also never been greedier. The twenty-six richest men in the world own more wealth than the poorest 50 percent of the world's population combined.[18] That's nearly 4 billion people. Greed is ultimately a form of lust, so the more you try to satiate it, the bigger it gets. As Winston Churchill said about Hitler when he first came to power, "His appetite may grow with eating."[19] Societies that censor or condemn greed, rather than tolerating or celebrating it, are less vulnerable to its toxic consequences—because they will have less of it.

If the bright side of vanity is corrupted by greed, it is completely extinguished in the absence of talent and hard work. In an age of rampant narcissism, fame and success are at the top of the values pyramid, irrespective of merit. Historically, our admiration for people, like their actual fame, was a product of their actual achievements and our acknowledgment that these resulted from some kind of merit rather than sheer luck of privilege. For example, we admired Maria Callas because of her amazing voice and presence, Velázquez because of his paradigm-shifting portraits, or Catherine the Great because of her vision and leadership. We have always admired famous individuals, but our tendency to admire individuals who are just famous for admiring themselves is a very recent phenomenon. In the AI age, if your fame is actually the result of certain talents or accomplishments, it almost seems second rate compared with those who managed to become famous for being famous. All style and no substance will get you further than no style and all substance. Same goes for politics and leadership: for every Angela Merkel, we have many Bolsonaros, Johnsons, Orbáns, and Putins.

So, for instance, Kim Kardashian's main achievement is to be famous without having any obvious talents, other than a sublime talent for self-promotion. The wide range of influencers you may never have heard of—like "Cooking with dog" (in which a Japanese woman explains how to cook traditional Japanese dishes while she is translated to English by her assistant dog); or "Ask a mortician" (which advocates for the reform of the Western funeral industry and covers a wide range of death-related matters, such as the current physical state of those who died in the *Titanic* and whether our nails continue to grow once we are dead). We have turned fame into a self-fulfilling prophecy, the result of gluing people's eyeballs onto some viral social media content and turning them into the main vehicle for influencing and productizing other people's behaviors with the help of ruthlessly efficient algorithms.

Since the AI age provides so many mechanisms for boosting our egos, we feel guilty as soon as we fail to achieve this. Every tweet or post you share with others is an attempt to bolster your reputation, and when you don't get instant positive reinforcement, it's as if you are being ignored or rejected by others.[20]

Unsurprisingly, research shows that people's depression and anxiety levels increase when they are unfriended on Facebook, which, let's face it, is as deep an approach to friendship as Martin Garrix is to the evolution of music, or Jordan Peterson to the evolution of philosophy (e.g., tidy your room, make friends with people who care about you, don't lie, and so on).[21]

Although the desire to be liked and accepted is a fundamental requirement in any society, and the basis for much pro-social behavior, if you are so worried about what others think of you, you can turn into a mindless conformist and lose any sense of independent or critical thinking. You'll also experience any form of negative feedback as apocalyptic and suffer from depression as soon as others reject you. This obsessional proclivity to depend on others has been equated to neurotic or insecure narcissism. We need others to inflate our egos in order to fulfill our narcissistic cravings. As soon as they fail to do this, our deep insecurities and inner vulnerabilities are exposed, and it feels unbearable.

The Authenticity Trap

Though over-depending on the opinions and approval of others is a big problem, the opposite is also true. If you don't care at all about what others think of you—if you choose to behave in spontaneous, unfiltered, and uninhibited ways—then you're going to act in selfish, toxic, and antisocial ways.[22]

Thinking of others can have pro-social effects. To use myself as an example (which is rather appropriate in a chapter devoted to narcissism), as I'm writing these words, I have to make a special effort to *avoid* being myself, taking into account what you, dear reader, may want to hear, what my publisher is interested in printing, and what other experts in this field may think of my ideas. All this is not a sign of weakness, mindless sheeplike conformity, or gullibility, but an essential ingredient for effective interpersonal functioning. My ability to relate to others in a healthy way fully depends on my willingness to conform to these expectations and norms.

Only in a highly narcissistic world would one of the most widely circulated pieces of popular advice on how to approach consequential career situations, such as a job interview, be the notion that the strongest formula for success is to "just be yourself" and not worry too much about what others think of you. This is probably one of the most harmful pieces of career advice ever given, and since people are still able to get jobs, one can only imagine that there is a high enough volume of people, if not a majority, that safely ignores this advice.[23]

In work settings, what people, and especially job interviewers, are interested in seeing is the *best* version of you.[24] That is, you on your best behavior, telling people what they want to hear, even if it isn't what you want to say impulsively. As the great sociologist Erving Goffman noted, "We are all just actors trying to control and manage our public image, we act based on how others might see us."[25] Adhering to the social etiquette, showing restraint and self-control, and playing the game of self-presentation will maximize your chances of landing a job, whereas being yourself may make you look spoiled, entitled, and narcissistic.

If you wanted to seriously adopt a free-spirit approach to life, you would have to go back to the very beginnings of life and avoid any influence from parents, family, friends, teachers, and culture altogether. In the unlikely event that you were to succeed at this enterprise, you would probably end up as a total outcast and misfit, marginalized from the very rules you have decided to ignore. Savant or primitive behaviors would be your normative modus operandi, and you could forget about language, manners, or adaptive functioning to any social setting. This would, in short, be the total opposite of what you normally do, which is to adopt socially acceptable or recommended behaviors and follow group-related norms.

A more moderate or loose interpretation of this mantra simply encourages us to release our social inhibitions, to approach each situation and especially important ones in an unfiltered, uncensored, and spontaneous way, as we would perhaps when in the company of close friends or relatives. So, for instance, if you go to a job interview, you may want to answer each question honestly. If you have a big client meeting, you may speak up openly and freely, even if it means revealing your true views about the product or indeed the client. And if your colleagues ask you whether you are happy to see them, you can tell them quite frankly that it pains you to have to interact with them, and so on. Equally, when you are going on a date, you can reveal your worst habits; after all, if they like you, they should like you as you truly are, with your authentic flaws. And others will admire or appreciate even your flaws so long as they are genuine, while your artificial virtues will have less value, for they don't represent our natural or real self. While this version of being yourself is more attainable, it is also more counterproductive.

As Jeffrey Pfeffer points out in his excellent book *Leadership BS*, being "authentic," in the sense of exhibiting one's true feelings, "is pretty much the opposite of what leaders must do."[26] This runs counter to most (bad) self-help advice, which encourages leaders to behave in spontaneous and uninhibited ways without much concern for what others may think of them and being truthful to themselves. Not only is this advice data-free, but it also probably accounts for a great deal of the problems leaders create in their teams and organizations. The best and most effective coaching focuses on helping leaders inhibit their spontaneous and authentic tendencies to instead develop an effective behavioral repertoire that replaces these natural or default habits with more considerate, pro-social, and controlled behaviors. In that sense, coaching is largely an attempt to dissuade leaders from expressing their authentic self. Why be yourself if you can be a better version of yourself? Why do what feels natural, when you can pause, think, and act in a way that makes you more effective? Nobody is a leader to be themselves, but to have the most positive influence on others that they can, which typically requires careful consideration, attention, and management of their actions, including repressing their natural instincts if needed.

For a species so accustomed to impression management and deception, it is perplexing how distasteful and immoral most people find the fact that the actual norm in any society—not just modern Los Angeles, Victorian England, or twentieth-century high-class Vienna—is not authenticity but to fake it, though such a reaction is itself testimony to the pervasive power of impression management, which we have internalized to the point of being unaware of it.

Even if we were actually incentivized to act in spontaneous and natural ways, it would not be easy to accomplish this. Perhaps this is the strongest evidence for the fact that we are naturally prewired to fake it, preprogrammed to restrain, adjust, and inhibit ourselves in any meaningful social setting since an early age. This is why parents shouldn't be surprised that from a very young age, their kids are generally well behaved when they visit other parents' kids, but not at home. Very early in life, we learn about the importance of *not* being ourselves, particularly in high-stakes situations, which is why if you are a well-adjusted and mature adult, you will find it rather difficult to follow the "just be yourself" prompt.

Meanwhile we spend a great deal of time curating our virtual selves in an effort to harness our digital identity in order to please others.[27] We pick photos we love, selectively report our success stories, while hiding our deepest anxieties and politely celebrating other people's equally fake and underwhelming achievements. As an old social media joke stated, nobody is as happy as they seem on Facebook, as successful as they seem on LinkedIn, or as smart as they seem on Twitter. But what if there is very little left in terms of behavioral remains or reputation after we accounted for those online activities and all the other digital platforms that consume our everyday life?

Needless to say, there is a comforting and reassuring aspect to the notion that there's more to us than what people see, which would also imply that AI can at best mimic our public or professional persona rather than our true or real self. Then again, like our work colleagues, AI may be less interested in our true self than the person who shows up every day, works, acts, and relates in a given way, regardless of what their inner self may be like.

Your true self may need ten years of psychotherapy to be discovered, but you can get to a pretty good model to predict your relevant everyday behaviors after a few weeks. Importantly, your true self is someone who perhaps four or five people in the world have learned to love—or at least tolerate—and not for the entire duration of the Christmas lunch.

Only Humility Can Save Us

If you accept the premise that our narcissistic culture is out of control, and that AI-based technologies are not only a huge beneficiary of this but also exacerbating it, then it makes sense that you wonder how on earth we can make things better, how can we escape this world of ubiquitous self-absorption, and what is the main antidote to the age of the self?

Humility, the ability to understand your limitations and avoid overestimating your talents, could be the answer.[28] At the individual level, being seen as humble is associated with a more positive reputation and higher degrees of likability. We think of someone as humble when they seem more talented than they think they are. Contrast this to narcissists who are *less* talented than they think they are, or at least want to think. Research shows that when we detect a surplus of hubris in people relative to their talents, they become less likable.[29] Perhaps if we remember this rule during our social media interactions, we can stop reinforcing narcissistic behaviors.

Another individual advantage of humility concerns personal risk management.[30] The less you bullshit yourself about your own talents, the more likely you will be to avoid unnecessary risks, mistakes, and failure. Only people who overestimate their abilities go to critical job interviews, client presentations, and academic exams ill-prepared. It also takes a certain level of arrogance to avoid medical advice, engage in self-destructive activities that put you and others in danger, and underestimate the risks of smoking, drinking, or drunk driving, or to refuse vaccinations during a global pandemic. By the same token, you are much more likely to develop new skills if you have accurately identified gaps between the skills you need and the ones you actually have.

There are collective advantages to humility, too. A society that values humility over narcissism will be less likely to end up with the wrong people in charge. In general, many narcissists are chosen as leaders because they're able to fool people into thinking that their confidence is a sign of competence. As the eminent historians Will and Ariel Durant wrote, "Human history is a brief spot in space, and its first lesson is modesty."[31] In my own research, summarized in my previous book, *Why Do So Many Incompetent Men Become Leaders?*, I highlight that the main explanation for the importance of humility in leadership is that we generally don't select for it.[32] Ironically, those who are self-deceived about their talents, to the point of being quite narcissistic, are often seen as leadership material. There is no better way to fool others if you have already managed to fool yourself. This comes at a cost, which is that we rarely end up with leaders who are aware of their limitations, and we often end up with leaders who are unjustifiably pleased with themselves, heroes in their own mind, and too arrogant to accept responsibility for their own mistakes or have awareness of their blind spots.

Having leaders who lack humility is particularly problematic during a crisis, as they will fail to pay attention to other people's feedback, take on board others' expertise, or be accountable for their poor decisions. To develop humility, you need to have some humility to begin with, and most people do. If, as a leader, you can accept your limitations, even if it's painful, you will have every opportunity to improve and get better. It is only by realizing that you are not as good as you want to be that you can embark on a genuine quest for self-improvement, which is what every leader must do. No matter how much potential or talent you have as a leader, you will need to get better to fulfill it. Great leaders are always a work in progress; a leader who is a finished product is probably finished. A simple thing you can do to develop your humility: pay more attention to negative feedback.

We live in a world that, fortunately, is quite civilized, but this also means that fake positive feedback, including ingratiation, is more common than candid critical feedback, particularly from your direct reports. And you need exactly that type of feedback. As my colleague Amy Edmonson and I have argued, one of the best indicators of humble leadership is to create a climate of psychological safety in your team and organization, where those who report to you feel free to provide you with negative feedback.[33] Try doing this by asking the right questions. Instead of saying, "Wasn't I great?" which is a way to encourage praise, ask "How could I have done this better?" "What would you have done differently?" "What is the single thing you would definitely want to change about my presentation, report, or decision?" And of course, be grateful when people give you this feedback, because it is not easy. They would find it much easier to suck up to you.

Humble societies—as well as groups, organizations, institutions, teams, and so on—would be far less likely to decay, since we'd devalue the traits that help narcissists thrive and we'd reward people for their actual talent and effort. In this kind of world, substance will trump style. And, rather than promoting those who think highly of themselves, we'll look for people who try to reduce their faults, problems, and imperfections, while remaining humble enough to want to strive for more. Last, but not least, a humble society would place more value on empathy, respect, and consideration than on selfishness and greed, because we would all agree that a fair system makes it easier for people to thrive on merit.

We all have the power to exercise more humility, no matter how much time we spend on TikTok. We also have the power to disincentivize others from acting in arrogant, self-important ways. Not falling in love with their own vanity, carefully scrutinizing their actual talents, and preferring others who are either humble or capable of faking humility in a convincing way are all fairly attainable ways to improve our cultural evolution in the AI age.

Humility is the possible cure for the malaise of arrogance and self-importance in the AI age. We may not be able to change our culture, but at least we can resist being influenced by it by rewarding humility rather than arrogance in others and behaving in humble ways ourselves.

TEST YOURSELF

Are You a Narcissist?

How narcissistic are you? Test yourself and find out.

- I love being the center of attention.
- I would rather be rich and famous than a good person.
- I am often jealous of other people's success.
- I am easily annoyed when others criticize me.
- People who know me appreciate my talents.
- I view myself more favorably than others do.
- I like to surround myself with people who admire me.
- I crave other people's approval.
- I am destined for greatness.
- I find it hard to fake humility.

Add one point for every statement you agree with; then add up your points.

0–3: You are a cultural outlier: the last humble human.

4–6: Consider yourself average; you can flex in the direction of more (or less) humility.

7–10: You are the perfect customer for social media platforms and a cultural referent for the AI age.

Chapter 6

The Rise of Predictable Machines

How AI turned us into very dull creatures

Maybe it's not until we experience machines that we appreciate the human. The inhumane has not only given us an appetite for the human; it's teaching us what it is.

—Brian Christian

Although AI has been rightly labeled a *prediction machine*, the most striking aspect of the AI age is that it is turning us humans into predictable machines.

When it comes to our everyday lives, algorithms have become more predictive *because* the AI age has confined our everyday lives to repetitive automaton-like behaviors. Press here, look there, drag down or up, focus and unfocus. By construing and constraining our range of behaviors, we have advanced the predictive accuracy of AI and reduced our complexity as a species.

Oddly, this critical aspect of AI has been largely overlooked: it has reduced the variety, richness, and range of our psychological experience to a rather limited repertoire of seemingly mindless, dull, and repetitive activities, such as staring at ourselves on a screen all day, telling work colleagues that they are on mute, or selecting the right emoji for our neighbor's funny cat pictures, all in the interest of allowing AI to predict us better.

As if it weren't enough to enhance the value of AI and Big Tech through our nonstop training of the algorithms, we are reducing the value (and meaning) of our own experience and existence by eliminating much of the intellectual complexity and creative depth that has historically characterized us.

Humanity as AI's Self-Fulfilling Prediction

When the algorithms *treat* or process the very data we use to make practical and consequential decisions and life choices, AI is not just predicting but also impacting our behavior, changing the way we act.

Some of the data-driven changes AI inflicts upon our life may seem trivial, like when you buy a book you'll never read or subscribe to a TV channel you never watch, while others can be life-changing, like when you swipe right on your future spouse. Most of us know of marriages that started (or ended) with Tinder, with research suggesting that more long-term relationships form via online dating than any other means.[1] Along those lines, when we use Waze to work out how to get from A to B, check the weather app before we get dressed, or use Vivino to crowdsource a wine rating, we are asking AI to act as our life concierge, reducing our need to think while attempting to increase our satisfaction with our choices.

Having trouble coming up with a new password? Don't worry—it will be auto-generated for you. Don't feel like finishing a sentence on email? All good—it will be written for you. And if you don't particularly feel like studying the map of a city you are visiting, learning the words of a foreign language, or observing the weather patterns, you can always rely on technology to do the job for you. Just as there's no need to learn and memorize people's phone numbers anymore, you can get by in any remote and novel destination by following Google Maps. Auto-translate AI will help you copy and paste any message to anyone in any language you want, or translate any language to yours, so why bother learning one? And, of course, the only way to avoid spending more time picking a movie than actually watching it is to follow the first thing Netflix recommends.

In this way, AI absolves us from the mental pain caused by having too many choices—something researchers call the *choice paradox*. With greater choice comes a greater inability to choose or be satisfied with our choices, and AI is largely an attempt to minimize complexities by making the choices for us. NYU professor and author of *Post Corona* Scott Galloway noted that what consumers want is not actually more choice but rather confidence in their choices. And the more choices you have, the less confident you will logically be in your ability to make the right choice. It seems that Henry Ford was exercising a strong form of customer-centricity when he famously stated that "customers are free to pick their car in any color, so long as it is black."

Regardless of how AI unfolds, it will probably continue to take more effort out of our choices. A future in which we ask Google what we should study, where we should work, or who we should marry is by no means far-fetched. Needless to say, humans don't have a brilliant track record at any of these choices, not least because historically we've approached such decisions in a serendipitous and impulsive way. So, much as in other areas of human behavior that become the target of AI automation—for example, autonomous vehicles, video interview bots, and jury decisions—the current bar is low. AI does not need to be too precise, let alone perfect, to provide value over average or typical human behaviors. In the case of self-driving cars, this would simply mean reducing the 1.35 million people who die in car accidents each year.[2] In the case of video interview bots, the goal would be to account for more than the 9 percent of variability in future job performance that traditional job interviews account for (correlation of 0.3 at best).[3] And in the case of jury decisions, it would simply require us to reduce the current 25 percent probability that juries have to wrongly convict an innocent person—that's nearly three in ten defendants.[4]

Thus, rather than worrying about the benefits of AI for effectively taking care of decisions that historically relied purely on our own thinking, we should perhaps wonder what exactly we are doing with the thinking time AI frees up. If AI frees us from boring, trivial, and even difficult decision-making, what do we do with this gained mental freedom? This was, after all, always the promise and hope of any technological revolution—standardize, automate, and outsource tasks to machines so that we can engage in higher-level intellectual or creative activities.

However, when what is automated is our thinking, our decision-making, and even our key life choices, what are we supposed to think about? There is not much evidence that the rise of AI has been leveraged in some way to elevate our curiosity or intellectual development, or that we are becoming any wiser. Our lives seem not just predicted but also dictated by AI. We live constrained by the algorithms that script our everyday moves, and we feel increasingly empty without them. You can't blame technology for trying to automate us, but we can, and should, blame ourselves for allowing it to squeeze any creativity, inventiveness, and ingenuity out of us just to turn us into more predictable creatures.

The problem is that we may be missing out on the chance to infuse some spark, richness, and randomness—not to mention humanity—into our lives. While we optimize our lives for AI, we dilute the breadth and depth of our experience as humans. AI has brought us too much optimization at the expense of improvisation. We appear to have surrendered our humanity to the algorithms, like a digital version of Stockholm syndrome. Our very identity and existence have been collapsed to the categories machines use to understand and predict our behavior, our whole character reduced to the things AI predicts about us.

Historically, we operated under the assumption that humans were psychologically complex and somewhat deep creatures, which is why it took us time to truly get to know someone. Say you are trying to help a fellow human understand who you are—a monumental, perhaps even impossible task. Consider that "who you are" could be broken down into the sum of all your behaviors—something difficult to track, record, attend to, and interpret, even in the current age of surveillance capitalism—*plus* the sum of all your thoughts and feelings, which is arguably even murkier. Moreover, consider all the levels of explanation needed to make sense of each aspect of you, from biological processes (e.g., your unique physiology, biology, genetic makeup, etc.) to social, psychological, cultural, and philosophical theories aimed at translating your personal patterns of activity into a meaningful model of you. At the low end of the complexity

spectrum is the AI approach to defining us: it substantially simpli-
fies the challenge and tackles the question in a very superficial, ge-
neric way, namely, to simply constrain the range of things you do,
feel, and think, limiting the potential repertoire of behaviors you are
likely to display on an average day, or lifetime, to improve whatever
mental models others have of you. In other words, rather than overly
simplifying the model, you can just try to simplify yourself.

The world of human creations may help illustrate this point. For
example, it is more complex to understand *Citizen Kane* than *Fast
and Furious 8*. It is more complex to understand Wagner than Ari-
ana Grande. It is more complex to understand Velázquez's *Las Meni-
nas* than some mass-produced elevator art hanging in the corridors
of a Hampton Inn. It's the same with people: we all differ in our
self-complexity, such that some of us are more "multidimensional"
than others.[5] Some people are so easy to read that even one or two
categories may help you understand and predict them, in the sense
that they are emblematic of their relevant categories. Some manage
to combine competing interests, antagonistic attitudes, and almost
paradoxical patterns of behavior; we may call them unpredictable
people. If you want to help others understand who you are, what
you are like, or what you are likely to do, just strive to eliminate any
complexity and unpredictability from your life. Reduce your life to
the obvious, the monotonous, and the repetitive, and my model of
you will rapidly grow in predictive power, no matter how simple.
For example, if my model of you is that of a human who will spend
their days looking at various screens and clicking away, tapping in,
scrolling down different pages in ever more repetitive fashion, even
a computer will be able to understand who you are. Acting like a
robot makes us more familiar to robots, and we are optimizing our
lives for this purpose.

Life on the Virtual Assembly Line

Does AI make us less free? Of course, free will is a big topic in philosophy, psychology, and neuroscience. Contrary to what common sense would dictate, scientists have the conventional view that there is no such thing as free will, that our decisions are made *for* rather than *by* us, and that consciousness is at best an inexplicable leftover of human evolution and at worst a made-up construct.[6] There is an interesting parallel to consider, namely, the role of predictability. When you can predict something, you have more control over it, and if you have more control over it, that renders that thing less free. Not that we can control the weather when we can predict it, but predicting it gives you more control over the weather and lessens the control that the weather has over you.

It is also philosophically complex to understand who is in control, say, when we act in predictable ways: an inner force that possesses us, the system, the AI, Big Tech, our nature, or just us. After all, I agree to watch the next YouTube clip, follow the suggested Waze route, or have another beer. Who is in control of my Uber ride—me, the driver, the Uber app, or some unknown cosmic force called the universe, God, or AI? Or do all these behaviors leave us guilty and helpless because we experience a lack of control and we rarely decide to alter them? In an insightful article, author Adam Grant recently noted that we are *languishing* rather than living, a word that could summarize many of the feelings evoked or at least imputed in this book.[7] Would a guilt-free life consist of controlling or inhibiting our tech temptations? And what precise aspect of it would actually make us feel free?

Many philosophers noted that it is OK to lack free will, so long as we can live under the illusion that we don't. The reverse, however, is far less palatable: feeling no control over our lives while we are actually free to shape and create them. In other words, equipped with free will but having no awareness of it whatsoever, as if we lived under the illusion that our choices are made for us and that we have been reduced to automatons or machines controlled by AI. This feeling of lacking agency not only signals a moral or spiritual defeat but also severely limits our sense of freedom and responsibility. Certainly, since AI has only recently arrived in our lives, there are limits to what we can blame it for. By the same token, there's a difference between controlling someone with data and simply collecting data on them. Much of what technology, and even AI, does is to monitor, measure, inspect. Many of the problems we attribute to AI or tech don't go away if we just shut down the algorithms, remove AI, or stop measuring things.

For example, my sleep patterns are no different when I disconnect my Oura Ring or forget to charge it. If, as philosopher Ludwig Wittgenstein noted, the difference between my hand going up and my raising my hand is free will, then the feeling of control or subjective experience of agentic action may be the key. That's what's we risk losing in an age where we lack analogue alternatives and feel easily trapped in a web of algorithmic predictions, irresistible but pointless digital nudges, and a never-ending, existential fear loop. Clearly, a Luddite life of recycled retro activities isn't precisely creative anyway, so how to create, what to create, and why?

Of course, the analogue world isn't free of these problems. Organizations have always sought to control and manage their workers through the design and structuring of tasks and activities around predictable, measurable, and improvable outcomes.[8] What began as "scientific management" with Frederick Taylor's assembly lines has now moved fully virtual in the AI age, exemplified by gig workers who are fully dependent on a platform and managed exclusively by algorithms. Taylor's idea was that workers would be controlled by a "distant brain," namely, senior management. Today that brain is AI, though it is still very much in development.[9] And before being alarmed by this, remember that senior management is not a very high bar to beat.

There are many potential advantages to having algorithms monitor, measure, and manage your performance, such as higher precision, objectivity, consistency, and the reduction of political favors and toxic behaviors.[10] For example, it's a lot harder to be harassed by AI than a human boss.[11] But there is also a clear dehumanizing side to being managed by a machine, particularly when its goal is to turn us into a machine, too.

Whereas previous technological breakthroughs, such as the industrial revolution, were about mechanizing manual labor—replacing human behaviors with machine activity—the current AI revolution, often called the fourth industrial revolution, is about mechanizing intellectual work, replacing human thinking, and learning with machine alternatives.[12] This, too, has positive sides. For example, self-driving cars will improve the lives of passengers and pedestrians because they will reduce the chances of dying or injury in car accidents and travel more safely from one place to another. Passengers can remain productive (or relax) while their cars drive themselves, not least because they have nothing else to do.

Our ingenuity has enabled us to increase efficiencies in every area of life. We have created technologies and automated systems that decrease the need for deep thinking. Where there is even a remote routine component to something, we have learned to economize thinking or reduced the need to think hard to just a single instance, so we can standardize decisions and responses to free up our minds for something else.

But to what end? If we're going to automate or outsource our judgment and decision-making, there are no strong indictors that we are willing to invest our released mental resources in any creative, inspiring, or enriching activities. As the brilliant author and professor Gianpiero Petriglieri points out, it is critical to defy the "dehumanization of workplaces," which can happen only if we stop optimizing for ever more efficiency and start reclaiming "refinement, rationality, and restraint."[13]

Sometimes, it seems as if we're all actors, performing the same role and reciting the same lines, night after night. When we work and live in a digital world, more and more deprived of proper analogue experiences, we are forced to remain constantly in role: we browse, click, and react; we forward, classify, and ignore. In the process, we risk ignoring life as it once was, simultaneously simpler and richer, slower and faster, serendipitous yet certain. Unsurprisingly, in recent years, we've seen a dramatic increase in mindfulness research and interventions, which aim to carve a bit of mental space, peace, and tranquility for our overly stimulated brains, enabling us to live in the moment and make a deeper connection with reality.[14] There is now a booming market for digital mindfulness apps, guides, and tools.[15] The irony is hard to miss, like a YouTube video on how to stop watching YouTube or the no-peeing section of a swimming pool.

The concerns about the degree of control that algorithms and AI have gained over us seem quite sensible and justified. If Amazon can predict what we buy, how much control do we have over it? If Netflix can predict what we want to watch, how much free will do we have? And if the world is cataloged around a vast range of metrics concerning what all of us and each of us will prefer, how much freedom do we have to behave in serendipitous ways? If we are the sum of the things AI can predict about us, do we still have the freedom to make unpredictable decisions? Or is our freedom only found in the decisions and behaviors that fall outside the algorithms' remit? Perhaps one day we will take pleasure in fooling AI, even with the simple act of deliberately misclassifying the wheels, lamp post, and traffic lights in a cybersecurity test, the common challenge deployed to identify whether we are humans or AI (by the way, I confess that I have personally failed this common "test of humanity" more than once, so I'm beginning to question my own status). Even if freedom is reduced to the perception of it—self-perceived free will, if you like—surely the way to increase it is to act in ways AI or the algorithms don't expect. A life optimized to refute AI and invalidate its predictions seems a life more interesting than one the algorithms fully account for. Want to join?

Engineering Serendipity

Serendipity used to be routine; it represented the basic syntax of life. We built it into our everyday existence and operating model as we learned to live with it, even when we didn't consciously love it. Still, there was something quite magical about realizing that someone you just met likes the same song or went to the same college, not to mention those random bar encounters, including those resulting in marriages. Now we look back to serendipity as a lost chapter in our lives. We don't totally miss it, but we wouldn't mind having some of it back. It is almost an acquired skill these days because our lives are predominantly optimized to avoid it.

How to change this? To begin, in the digital world, creativity and serendipity will come from acting in ways that algorithms can't expect or predict. For example, what will you choose to search on Google? The creative answer is something that Google autocomplete cannot predict. Equally, imagine saying something unexpected on Twitter, or watching something you don't usually watch on Netflix, and so on. We live in a world optimized for prediction, so algorithms (and their hosting platforms) are trying to standardize our habits to increase their predictive power. It's just like working with someone who is a bit obsessional or being married to a control freak: move one thing out of place and their models and systems will collapse. They feel anxious and out of context if you surprise them, so they try to make the world as predictable as possible. The same is true for Facebook or Google algorithms.

Broadly speaking, the goal should be to create a richer, more comprehensive, varied, diverse, and heterogeneous version of ourselves. Think of it as diversity at an *individual* level: to broaden our identity in order to encompass many *me's*, a multi-me, as entrepreneur Riccarda Zezza notes, that actually enriches our character and perspectives.[16] As a result, we would also become less predictable versions of ourselves, because our habits, beliefs, and opinions would not reflect a uniform ideology or narrow self-concept. When knowing a single thing about someone (e.g., whom they vote for, where they live, what they do, what their favorite food is, or what news channel they watch) is enough to accurately predict everything else about them, then there is not much complexity to them. Even intellectually, their consistent and narrow self-concept has saved them from having to think, reason, or make novel decisions. And the more predictable they are, the less control, agency, and creativity they have.

We are the hardest part of the world to change, but also the most important one. We need to examine our actions and create new patterns of behavior, reexamine our beliefs, and instill a minimum dose of innovation in our lives. If we don't, then we better resign ourselves to being sheer passengers or spectators in this world.

Despite all the money, power, and science behind the attempt to predict everything we do and understand everything we are, there are no obvious signs that AI has managed to truly decode us. So far, there is no evidence that either AI or any other tool can predict broad outcomes, such as our relationships, career, and general life success, with a great deal of accuracy. We are more than algorithms assume and are capable of unpredictable actions.

Randomness is in the eye of the beholder. If you create a situation in which people experience randomness and serendipity, but everything is staged and predetermined, you in effect recreate the free-will illusion: Who cares if I'm not free to choose so long as I *feel* that I am? Interestingly, AI often follows the reverse approach: it attempts to force predictability even when we have free will, mostly to oversell its predictive powers to marketers and the market. So, when an algorithm tells us that we probably want to watch X movie or read X book, there is something of a self-fulfilling prophecy here. If we trust the algorithms more than ourselves or have limited knowledge to evaluate the accuracy of such predictions, then we may just go for it, which may in turn reinforce the perceived accuracy of the algorithms. This placebo effect is reminiscent of the famous anecdote about Niels Bohr, the Nobel Prize–winning physicist, who allegedly kept a good-luck horseshoe in his office because he was "told it works whether you believe in it or not."[17]

There is more complexity to human behavior than AI can handle. We can create and shape our actions in ways the algorithms don't expect. The capacity to elude algorithmic predictions, the foundational arithmetic of AI, still remains a fundamental part of human creativity and freedom. This can be an asset in the future: our willingness to escape the little boxes or cages in which AI is trying to place us. This is a critical component of creativity: when you think about someone—friend, colleague, person—whom you designate as "creative," that is probably because they have the ability to surprise you, which means they disprove your predictions of their behavior. They render your models flawed or invalid.

Let's try to be less predictable. When you are told that you have certain habits that you don't like, it triggers certain desires to change and get better. If we were shown the models AI has for us and, in effect, watched a movie based on our everyday behaviors and tendencies, realizing how boring our lives have become, we could surely find a reason to try to become a more creative and unpredictable version of ourselves.

Watch the movies Netflix would never expect you to watch. Connect with people that Facebook or LinkedIn never expect you to connect with. Watch the videos YouTube never expects you to watch. Invalidating AI's predictions may be the ultimate way to harness a life beyond the algorithms' models of us or at least feel free.

One day in the not-so-distant future, we will be grateful for all the undocumented and unregistered moments of our lives that exist only in our memories. The experience of trying to share your memories with others who shared those actual experiences still surpasses anything you can see on YouTube. What is a friendship but the ability to recap shared experiences with others, especially when there's no digital records of them? And as most of us will have experienced during the Covid-19 pandemic, digital friendships are not quite the substitute for face-to-face friendships, which provide a much wider range of psychological benefits.[18] Likewise, despite attempts to create "friendly AI," these consist mainly of polite bots that, much like dogs, are friendly in the sense that they remain nice and docile even when mistreated.[19]

Perhaps when our kids grow up, they will meet their childhood friends and ask, "Do you remember when I liked your TikTok video?" or "Do you recall when we watched that Instagram feed?" or "Do you remember the first time we saw KSI play FIFA on YouTube?" but it seems those memories will be somewhat less intense than what happens when you are not on TikTok or Instagram.

If the ability to thrive as humans largely depends on finding space, then the challenge in this AI chapter of our cultural evolution is to find or make space between the algorithms, our phones, screen time, and so on. When Viktor Frankl recalls his concentration camp experiences, he makes it clear that humans have a unique ability to create space out of nothing and express themselves even in the most constrained and inhumane situations.[20] In a world where everything relevant to us is increasingly digital, then the irrelevant things may be what makes being human an interesting endeavor.

AI, as of now, does not pass the Turing test when it comes to creativity. But it is sure to get better. AI for music creation is already pretty mature, with several platforms and programs that offer plug-and-play improvisation and composition, based on whatever parameters we pick, including emotional sentiment and style. Most of the tech giants, including Microsoft, Google, IBM Watson, and Sony, as well as specialized startups like Aiva and Amper, have off-the-shelf products, which have been used to create commercial albums, such as *I Am AI* by YouTube star Taryn Southern.[21]

To give you a sense of how far advanced this area is, AI—running on a smartphone—has been used to complete Schubert's famous Symphony No. 8, which the composer had abandoned after two movements. Stanford researchers have trained AI to effectively improvise jazz in the style of Miles Davis (with a 90 percent validation rate from human listeners).[22] There are many more examples. You can imagine that things are much easier with less complex genres than classical and jazz, not least because even before AI was used to compose music, we had elevator music that sounded like a total machine creation. AI has already written many books with dedicated online stores, such as www.booksby.ai.[23] For example, Google's AI ingested eleven thousand books to create poems like "horses are to buy any groceries. horses are to buy any animal. horses the favorite any animal."[24] Despite the tough job market for poets, this suggests that they are not about to be automated.

The problem with judging AI's artistic abilities is that art is a very subjective thing. For instance, the main difference between Joseph Beuys's *The End of the Twentieth Century* and a pile of large stones, like the difference between Tracey Emin's *My Bed* and my bed, is that neither my bed nor any other large stones found their way into the Tate Modern. This also means there is a big difference in value or valuation, which of course is entirely subjective or culturally determined. So, when Christie's recently sold an AI-generated work called *Portrait of Edmond de Belamy* for $432,500, there was nothing new about those who objected to this being real art. In reality, algorithmic art has existed since the early seventies, so perhaps the main change is that the quality of artwork got better or that we are now more open to designating AI creations as "art," or both.

As AI advances, the only thing that sets humans apart is that we can do a bit of everything, even if AI outperforms us on each of those tasks. It is the inability to develop artificial general intelligence that still makes general human intelligence quite valuable, and the ability to act in unpredictable ways that limits AI's scope to accrue human-like abilities.

This is why experts such as Margaret Boden, a professor of cognitive science, have argued that the main learning from the AI age so far is that the human mind is a great deal richer, more complex, and subtle than we originally assumed, at least when the AI field got started.[25] Perhaps this is the clue to maintaining our relevance or even our edge over machines, demonstrating our creativity in the future: surprising not just other humans but also AI. Surprise is a fundamental feature of creativity. If you are not acting in unexpected or unpredictable ways, then you are probably not creative.

You need to defy the order of things, go against the status quo and convention, and think that you are able to make things better. You have to disdain to some degree the ideas that are right, because creativity is never right in expected ways. Poet Charles Bukowski said, "Find a passion and let it kill you." That is a good illustration of the creative process, or what philosopher Martin Heidegger described as "being-towards-death."

Importantly, there are many practical ways to try to boost your creativity. For example, better prioritization (which includes saying no more often), changes in routine (such as taking a different route to work, adopting new habits, exposing yourself to new people, ideas, and topics), freeing time to unleash your curiosity, and finding hobbies or activities that make you *feel* creative (e.g., cooking, writing, playing music, photography, designing something) may all produce rich and valuable deviations from your predictable life, to the chagrin of AI.[26]

Are You Predictable?

Ideally, the question whether you have become or are at least acting like a predictable machine would be asked of other people rather than you. Otherwise, it feels like asking AI or a chatbot whether it is human. However, the following statements are designed to help you get a sense of whether you are living a standardized life that, at least from a subjective experience perspective, feels somewhat robotic.

- I am not totally sure that I haven't been automated.
- At times, I feel like I have little freedom or control over my behaviors.
- Everything I do seems to be predictable and repetitive.
- I can't remember the last time I surprised myself by doing something different.
- My life feels like *Groundhog Day*.
- I feel like the algorithms that mine my life know me better than I know myself.
- Time goes by very quickly.
- I often experience a sense of languishing.
- I don't feel like I have the freedom to be creative anymore.
- I often feel like a robot.

Add one point for every statement you agree with; then add up your points.

0–3: You are probably not spending as much time connected as your peers, or somehow find ways to surprise yourself and unleash your creativity.

4–6: You are average and within a range that can easily be flexed into the direction of more creative or more robotic behaviors. You may not have become a predictable machine yet but may want to try to remain somewhat spontaneous and free.

7–10: You are the perfect customer of social media platforms and a bastion of the AI age. Now for the good news: being aware of these issues is necessary for achieving any positive change. When would you like to start?

Chapter 7

Automating Curiosity

How the AI age inhibits our hungry minds

Nothing in life is to be feared, it is only to be understood.

—**Marie Curie**

Among the wide range of thinking skills that make us the humans we are, few have been made more important by the AI age than curiosity, the desire or will to learn.

All humans are born with an instinct for curiosity; studies have measured curiosity in four-month-old babies.[1] If you've spent any time around infants, you've seen how different things capture their attention, and *looking time* is one of the earliest indicators of infant curiosity: the more time kids spend looking at something, particularly novel objects, and the more interest they show in novel stimuli, the more likely it is that they will display curiosity as adults.[2] In other words, the hungry mind of adults starts developing from a very early age.[3] Sadly, it is also true that curiosity levels tend to peak at the age of four or five for most humans, only to decline afterward.

Even though past civilizations, from Ancient Egypt and Greece to the French and British Enlightenment, harnessed a reputation for open-mindedness and curiosity, at least relative to others, civilization has generally conspired to hinder or tame our hungry minds. Throughout much of our evolutionary history, there were generally limited incentives for displaying curiosity.

Instead, our main survival tactics favored only a minimal dose of curiosity, such as for exploring new sources of food or trying out new hunting tools.[4] Consider hunter-gatherers. If tribe X were to wander around the African savanna and one of its members suddenly felt the need to wander in a different direction to explore what tribe Z was doing, perhaps in the hopes of mitigating his boredom or overcoming his unconscious biases, he would be killed and eaten. Best-case scenario? Return to tribe X with interesting stories, but also more parasites, which might extinguish many of his peers. As we all learned during the Covid pandemic, the risk of parasitic infection is significantly higher if you socialize outside your closed circle.[5] That may mean twenty people in Italy and two in Finland, but there is always a tax on exceeding your biological safety zone, especially if you want to keep parasites at bay.

Is AI More Curious Than Humans?

Although the suppression of curiosity is rooted in our evolutionary history, technology has exacerbated it. To some degree, most of the complex tasks that AI has automated are evidence for the limited value of human curiosity over and above targeted machine learning. Even if we don't like to describe AI learning in terms of curiosity, it is clear that AI is increasingly a substitute for tasks that hitherto required a great deal of human curiosity, from finding the answer to all the pressing questions we now Google to exploring potential careers, hobbies, vacation destinations, and romantic partners.

Most AI problems involve defining an objective or goal that becomes the computer's number one priority. To appreciate the force of this motivation, just imagine if your desire to learn something ranked highest among all your motivational priorities, above your need for social status or even your physiological needs. In that sense, AI is way more obsessed with learning than humans are. At the same time, AI is constrained in what it can learn. Its focus and scope are very narrow compared with that of a human, and its insatiable learning appetite applies only to extrinsic directives—*learn X, Y, or Z*. This starkly contrasts to AI's inability to self-direct or be intrinsically curious.[6] In this sense, artificial curiosity is the exact opposite of human curiosity; people are rarely curious about something because they are told to be. Yet, this is arguably the biggest downside to human curiosity: it is free-flowing and capricious, so we cannot just boost it at will, either in ourselves or in others.

Innovative ideas once required curiosity, followed by design and testing in a lab. Now, computers can assist curiosity efforts by searching for design optimizations on their own. Consider the curiosity that went into automobile safety innovation, for example. Remember automobile crash tests? Due to the dramatic increase in power, a computer can now simulate a car crash.[7] With this intelligent design process, the computer owns the entire life cycle of idea creation, testing, and validation. The final designs, if given enough flexibility, can often surpass what's humanly possible.

Consider a human playing a computer game. Many games start out with repeated trial and error, so humans must attempt new things and innovate to succeed in the game: "If I try this, then what? What if I go here?" Early versions of game robots were not very capable; they knew where their human rivals were and what they were doing, but that didn't make them better than them. But since 2015, something new has happened: computers can beat us on equal grounds, without much contextual information, thanks to deep learning.[8] Both humans and the computers can make real-time decisions about their next move. (As an example, see the video of a deep network learning to play the game *Super Mario World*.[9])

With AI, the process of design-test-feedback can happen in milliseconds instead of weeks. In the future, the tunable design parameters and speed will only increase, thus broadening our possible applications for human-inspired design.

Consider the face-to-face interview, which every working adult has had to endure. Improving the quality of hires is a constant goal for companies, but how do you do it? A human recruiter's curiosity could inspire them to vary future interviews by question or duration. In this case, the process for testing new questions and grading criteria is limited by the number of candidates and observations. In some cases, a company may lack the applicant volume to do any meaningful studies to perfect its interview process. But machine learning can be applied directly to recorded video interviews, and the learning-feedback process can be tested in seconds.[10] Interviewers can compare candidates based on features related to speech and social behavior.

Micro competencies that matter—such as attention, friendliness, and achievement-based language—can be tested and validated from video, audio, and language in minutes, while controlling for irrelevant variables and eliminating the effects of unconscious (and conscious) biases. In contrast, human interviewers are often not curious enough to ask candidates important questions, or they are curious about the wrong things, so they end up paying attention to irrelevant factors and making unfair decisions. Yet research shows that candidates still prefer face-to-face interviews, in a clear case of "better the devil you know."[11] Perhaps the prospect of being liked by humans matters more than the ability to impress the algorithms— or is it that we prefer to be discriminated by humans rather than AI?

Computers can constantly learn and test ideas faster than we can, so long as they have clear instructions and a clearly defined goal. However, computers still lack the ability to venture into new problem domains and connect analogous problems, perhaps because of their inability to relate unrelated experiences. For instance, the hiring algorithms can't play checkers, and the car design algorithms can't play computer games. In short, when it comes to performance, AI will have an edge over humans in a growing number of tasks, but the capacity to remain capriciously curious about anything, including random things, and pursue one's interest with passion may remain exclusively human.

Even if we train AI to display actions that resemble human curiosity, for every problem AI solves, there will be many questions that arise. The questions are generally not asked or formulated by AI, particularly if we are talking about novel questions, but by humans. A good way to not just think about AI, but also leverage it, is to have it answer the questions we generate, which should really leave us more time to ask questions that can be answered and hypotheses that can be tested.

Wait, But Why?

Throughout human history, a variety of thinkers have pointed out that the most salient aspect of knowledge or expertise consists not in finding answers, but rather in asking the right questions. Most notably, Socrates believed that the role of the philosopher is to approach any conversation with an inquisitive mind to extract answers to even the most profound questions from ordinary people, so long as you can guide them with evermore poignant questions.[12]

We can imagine a future in which AI has all the answers, even if it got there by crowdsourcing them from all people and we are left with the main responsibility of asking relevant questions. Google would have been a dream to Socrates, since it can give us all the answers, so long as we ask the right questions. And it gets smarter the more questions we ask. We, on the other hand, get dumber, because our motivation to learn, to retain facts, to dig deeper and go beyond the surface of information that we are served with the first hit in a Google search decreases, and with it our intellectual curiosity.

Google's AI can answer most of the questions we ask (apparently only 15 percent of questions asked have never been asked before).[13] But it cannot yet help us ask the right questions; the closest it gets is to expose the similarities between the sentences we start when we ask questions and other sentences asked by most Google users deemed similar to us, at least in terms of current location.

But perhaps human curiosity has not yet been deemed worthy of automation because it would be like AI disrupting or cannibalizing itself. Google sells answers to our questions, whether smart or dumb. If it killed our questions, it would be left with nothing to sell, unless it was truly able to predict what questions we wanted to ask and managed to monetize or sell its answers without having us go through the trouble of asking the questions in the first place.

Automating certain aspects of curiosity is an interesting and useful objective for AI, and not incompatible with our incentive to free up some of the basic and predictable elements of our curiosity—like knowing where the bathroom is—in the interests of asking deeper questions. The only way to automate, and perhaps even kill, curiosity is to answer every question we have or convey the illusion that all our questions, both actual and potential, have been answered. This is why we have little incentive to waste time on asking questions or thinking much about any answers: it can all be Googled anyway and for everything else there is Mastercard.

Think of a near-term future in which we just ask Google whom we should date, marry, or work for. After all, with the volume and depth of data Google has on us, and its continuous refinement of its algorithms and neural networks, you can expect Google's answers to these questions to be less imprecise and certainly more data-driven than our own intuitive—and subjective—choices, not to mention our parents', friends', or crazy uncle's.

Reclaiming Curiosity

The fact that curiosity is becoming rarer in the AI age explains why it is in high demand. Indeed, curiosity is hailed as one of the most critical competencies for the modern workplace, and research evidence suggests that it is not just a significant predictor of an individual's employability—a person's ability to gain and maintain a desirable job.[14] Countries where people have higher curiosity levels also enjoy more economic and political freedom as well as higher GDPs.[15]

As future jobs become less predictable, more organizations will hire individuals on the basis of what they *could* learn, rather than on what they already know. Of course, people's careers still largely depend on their academic achievements, which are (still) influenced by their curiosity (more on this later).[16] Since no skill can be learned without a minimum level of interest, curiosity may be considered one of the critical foundations of talent.[17] As Albert Einstein famously noted, "I have no special talent. I am only passionately curious."[18]

The AI age has exacerbated the importance of curiosity with regard to job and careers. For each job that AI and technology automate, new jobs are created, but they require a new range of skills and abilities, which in turn will require curiosity or what some have referred to as *learnability*, the desire to adapt your skill set to remain employable throughout your working life.[19]

At a World Economic Forum meeting in Davos, Manpower-Group predicted that *learnability* would be a key antidote to automation.[20] Those who are more willing and able to upskill and develop new expertise are less likely to be automated. The wider the range of skills and abilities you acquire, the more relevant you will remain in the workplace. Conversely, if you're focused on optimizing your performance, your job will eventually consist of repetitive and standardized actions that a machine could better execute.

Being curious—or open-minded—is easier said than done. Within psychology, there is a rich history of studying open-mindedness, usually under the label of "openness to experience." Perhaps because the data is heavily skewed toward US psychology students, who are generally as liberal as the academics who conduct this research, open-mindedness has more or less been defined as politically liberal or, shall we say, Democrat rather than Republican. If this were exclusively a measure of political orientation, the only questionable thing would be the label, which celebrates one political affiliation while condemning the other. However, matters are made worst by the fact that we have glorified openness as

a general measure of curiosity, artistic inclination, cultural refinement, and verbal intelligence. So, people with high openness scores
are more liberal, less religious, and more intellectually oriented. As
you can imagine, this means they usually don't mix with conservative, religious, or culturally unsophisticated individuals. And, of
course, the feeling is mutual.

A truly open human being, however, would not fall into one or
the other extreme, but oscillate slightly within the middle range of
scores. They would, to use the now infamous Cambridge Analytica
term, be psychologically *persuadable*, for their ideology would not
preclude them from trying to connect with those who are different or motivate them to exclude someone from their circles based
purely on the fact that they don't agree with them or have different
lifestyles, backgrounds, and so on. This kind of open-minded, non-
partisan, value-agnostic thinking is not just hard to achieve; there
are actually few incentives to do so. In theory, it sounds like a great
idea and logically critical to increase diversity and inclusion in any
group or organization. The reality is that it would be intellectually demanding—requiring us to evaluate everything and everyone
with a clean slate, therefore ditching our ability to *think fast*—and
marginalize us from our friends and social circles. As society becomes more tribal, there are fewer rewards for those who behave in
anti-tribal ways.

Much of today's debate about *cancel culture* is not so much driven
by open-minded academics, but by conservative ideologues who feel
that the left has hijacked the narrative in academic circles. They have
a point: universities have become psychologically homogeneous,
and in top US colleges, Democratic professors outnumber Republicans by a factor of nine to one.[21] Although there is a fair amount
of self-victimizing by a few controversial right-wing thinkers who

have been uninvited from campuses to give their contrarian lectures, there is clearly logic to this argument: universities should help people think for themselves and nurture a critical mind, which is impeded by exposing them only to people who think like them.

What does this mean for current diversity and inclusion programs, which are basically asking people to accept, and ideally like, those who are totally different from them? It's like hoping to erase three hundred thousand years of evolution, just to ensure that our employer doesn't get into trouble and is seen as a champion of justice and equality. Of course, that is not how many people see it, for much of humanity is truly committed to increasing meritocracy and fairness, and overriding millennia of apelike attitudes toward *the other*. Culture has the power to influence how our primordial instincts are manifested, but it does not silence our genes. The only sane and honest conclusion one can make about humans and diversity is that the brain is biased by design. This means the best we can do is to eliminate cultural sanctions to our curiosity and praise people for exploring novel and distinct environments, including other people.

While we love to tell others that we are mysteriously complex and unpredictable, we would be the first ones to freak out if we truly were unpredictable: imagine seeing a different person every time you look at yourself in the mirror, or having no sense of what you would do in a given situation (e.g., meeting a client, going on a date, or getting your Starbucks coffee). We need to make sense of our actions, interpret, and label our motives, until we organize a well-rounded and meaningful self-concept that easily conveys who we are, to others and to ourselves. We need to stitch together all the tiny behavioral fragments in our everyday life to paint a cogent self-portrait. In a raw format, we are profoundly decomposed, so our job is to reconstitute, recompose, and reformat.

Putting labels of categories on others to summarize their entire existence is easy, but doing this with ourselves is hard, which is why we don't accept it when others do it about us. The fact that you cannot agree or disagree with anyone on either side without being (successfully) pigeonholed into one camp or another shows how predictable our attitudes and identities have become. Tell me one thing about you and I will tell you everything else. Consider the basic difference between a Trump voter and a Biden voter; it is all you need to know about someone to predict whether they love or hate guns, vegans, climate change activists, and gender-neutral pronouns. These snippets of identity are meaningful indicators of people's values, a rapid guide to sensible decision-making or a moral compass that organizes their actions with the goal of making them feel rational, decent, and predictable. We all create and love our filter bubbles, but that is how we become less curious and more narrow-minded as we age.

Another simple hack is to develop the habit of asking *why*, which is something we are all good at as three-year-olds but struggle with in our thirties. And another one is to impose regular routine changes in our lives so that we increase our own exploratory behaviors and add more variety and unpredictability to our lives. People tend to make the world as predictable and familiar as possible so it doesn't freak them out, but that's also how we end up mentally lazy.

Evolving Our Own Intelligence

Irrespective of AI's ability to advance, it seems clear that in the AI age, the essence of *human* intelligence is highly conflated with humility and curiosity, perhaps more so than with knowledge or logic. When all of the world's knowledge has been organized, stored, and is easily retrievable, what matters most is the ability to ask questions, particularly the right questions, and the willingness to evaluate the quality of the answers we get. We also need the humility to question our own intelligence and expertise so that we can remain grounded and maintain an inherent desire to learn and get better.

So, if we take AI as the context for human intelligence, what does intelligence need to do differently in the presence of AI? As Ajay Agrawal and colleagues note in *Prediction Machines,* since AI is basically a computer engine devoted to the rapid and scalable identification of patterns, we are better off putting our intelligence to a different use.[22] After all, we can expect machines to spot co-variations, co-occurrences, and sequences in data sets with a level of precision that far exceeds even the most advanced of human capabilities and the smartest human. When AI masters and monopolizes the task of prediction, the fundamental role of human intelligence is confined to two specific tasks: (1) structuring problems as prediction problems; (2) working out what to do with a prediction.

The first is in essence what scientists have been doing for centuries: to formulate testable—falsifiable—hypotheses and define observations that can help them test or support their assumptions. An example of the first task could be "if we spend less time in meetings, we will become more productive," meaning that average meeting time could indicate or signal performance. An example of the second task could be "since this manager spends 40 percent more time in meetings than others, I will tell her to spend less time in meetings" ("and if she doesn't, I will replace her with someone else").

Although these principles may sound obvious, there is one major obstacle to putting them in place: our own intuition. There will be many times when AI will run counter to our common sense, telling us to do A when our gut tells us to do B. Just as when Waze or Google Maps recommends that we follow route A, but our own experience and intuition tell us to take B, the data-driven insights derived from any AI may conflict with our own instincts. If you think of yourself as a savvy driver, and you feel that you know your way around a town or city, you may still use Waze but decide to trust it only on occasions—when it confirms your own instincts.

When it comes to structuring problems as if-then scenarios, the issue is that there are so many potential variables and factors to consider. We tend to select those that interest us most, which already captures many of the biases governing the human mind. Consider the manager who designs a highly structured job interview, putting candidates through the exact same questions and with a predefined scoring key to interpret and score their answers against relevant job demands and requirements. Although this type of interview has long been identified as one of the most accurate methods for selecting employees for a job, it is not without bias.[23] One of the hardest ones to address—at least for humans—is the fact that any a priori selection of if-then scenarios will be limited, so even before the interview starts, interviewers will have already cherry-picked certain attributes and signals while ignoring others. The more comprehensive and thorough they try to be, the harder it is for them to pick up the right signals during the actual interview—we can really only pay attention to one thing at a time.

In theory, it is sensible to think of structuring problems as prediction problems. But in practice, we have limited capacity for laying out multiple predictions at once, let alone paying attention to the outcomes. Our imaginary hiring manager may well have identified important indicators they need to observe in their interview candidates, but they will probably represent a small subset of the potential patterns—both relevant and irrelevant—that emerge in the actual interviews. For instance, if candidates speak a lot, they may be more likely to be narcissistic or self-centered (if-then pattern), so I shall not hire people who speak a lot (what to do with a prediction). But this pattern may just be one of hundreds or even thousands of patterns that increase the probability of someone being a misfit for a job. And even if there was just one pattern, that doesn't mean that we are good at spotting it, at least not in a reliable way.[24]

Let us assume that our hiring manager decides that someone is a good candidate for a job. Can we really be sure that they truly decided based on their logic, rather than on some other factors? Perhaps on paper the person met the formal criteria—like the person didn't speak too much—but what about all the potential reasons that were not registered or accounted for? These would include the candidate's charisma, likability, and attractiveness.[25]

The same applies when we judge famous artists or actors on their talents. For instance, I always found Cameron Diaz to be a good actress, as well as funny and charismatic, but perhaps I am just attracted to her. And when someone persuades himself that their colleague is boring or unintelligent, can we be sure that this decision is based on facts rather than the colleague's race, gender, class, or nationality? The science of consciousness is complex and inconclusive, but one of the little-known facts is that free will is an illusion, for we make most decisions influenced by an array of neurochemical activity that is affected by tons of factors other than logic, from the amount of sunlight to room temperature, sleep quality, caffeine consumption, and, of course, a firm preference for being right rather than wrong, one who is rarely intimidated by facts.

Our Deep Desire to Understand

Even if AI is capable of acquiring something along the lines of human curiosity, most AI applications to date have been in the prediction phase, with very few instances when prediction translates into understanding—that is, going from *what* to *why* depends on human intelligence and curiosity. It is our expertise and insights that turn predictions into explanations, and it is our own deep desire to understand things that differs from AI's relentless obsession for predicting things. We may well be the sum of everything we do, but the mere sum of what we do is insufficient to explain who we really are or why we do what we do. In that sense, AI's curiosity is rather limited, at least compared with humans. Even when we are unable to predict what we or others do, we have the capacity to wonder. Though there would be astonishing merit in deciphering who we are, particularly if computers could do this in a seamless, scalable, and automated way, the more mundane reality is that the big AI firms have gotten fat and rich without even trying.

We don't always buy what Amazon recommends, watch what Netflix suggests, or listen to what Spotify picks for us (though, apparently, I do). Part of the reason is that we still trust our instincts more than AI, even though we allow our instincts to become more data-driven and AI-guided. The other part is that we don't always *understand* the reason for those predictions. It's as if an online

dating app suggested we marry Yvonne, because that's what AI recommends, without actually explaining the reasons. It would be a
lot easier to trust AI if it didn't just stop at the prediction stage but
also included an explanation: marry Yvonne because you and she
share similar interests and will have great conversations, wonderful
sex, smart children, and so on. By the same token, if AI recommended our best possible job, we would want to know why that's
a good fit for us: Is it the company culture, the people, the nature
of the role, the potential to grow and get promoted, or is it that all
other options are just worse? In short, we want AI to not just predict but also explain things.

So far, most of the behavioral recommendations of AI can rely on
data-driven insights that humans can leverage to make decisions,
unless they decide to ignore them. AI has little need for theory; it
is a blind or black-box data-mining approach. In contrast, science
is data plus theory, and it still represents the best bet for acquiring knowledge because of its replicable, transparent, ethical, and
explainable approach to formulating and evaluating hypotheses. If
we accept the basic premise that AI and human intelligence share
a quest to identify patterns or to link different variables with each
other to spot covariations, then we must acknowledge the potential AI has for advancing our understanding of people, including
ourselves.

Clearly, Spotify is capable of predicting my musical preferences
very well, better than my best friend. However, that is not a convincing indicator that it actually knows me in the sense that my
best friend does, let alone understands me. To actually understand me, Spotify would need to add context, theory, and more
data to its models, data that goes beyond my musical preferences.
This is why my best friend knows me better than Spotify, because

she has a far broader range of behavioral signals to draw from, context, and a theory, albeit rudimentary, to translate this data into insights. My friend may not have a 100 percent or even 60 percent hit rate recommending music to me, but she may immediately realize that my choice of songs today reveals I'm feeling nostalgic, more so than usual. She may also understand why: Diego Maradona, the greatest footballer of all time and Argentina's finest export, has died. And she, of course, can share my sadness at this event and understand that our devotion for this epic player and iconic cultural figure also indicates certain anti-establishment, contrarian, rebellious aspects of our personalities, which we don't just share but which also explain our friendship to begin with—and as of today, you need to be human to understand this sentence.

One of the nice things about friends is that they understand us. Friendships are valued because we have an inherent need to be understood. As the famous quote from Orwell's *1984* states, "Perhaps one did not want to be loved so much as to be understood."[26] The other nice thing about friends is that we seem to understand them, which highlights the human need to make sense of the world. Our close friends may be one of the few things in the world we actually understand. It is precisely this understanding that provides the foundations of shared experiences, camaraderie, and affection that make human relationships better than human-machine relations, so much so that we can still enjoy connecting to humans even when it happens via machines. It's a formula that cannot be scaled or applied to everyone, but there are clearly good reasons for trying to replicate this deep understanding to the wider world. It would be nice to live in a world where people are understood, which would also require us to understand ourselves.

Our world is very much the opposite: it suffers from too little understanding or a crisis of misunderstanding. As a consequence, people make irrational decisions on consequential matters such as careers, relationships, health, and life in general. As anthropologist Ashley Montagu noted, "Human beings are the only creatures who are able to behave irrationally in the name of reason."

The bad news (or perhaps the good news) is that we are still a far cry from having "empathetic" AI, that is, AI that actually understands how we think and feel, let alone cares about it. Even when AI can predict our choices and preferences better than our friends can, it has no working model of our personality. In that sense, the difference between AI and human intelligence is largely based on people's ability to understand humans. If I told you that every night Jake takes out $400 from an ATM outside the casino, your conclusion will be that Jake has a gambling problem (and a lot of money to burn). On the other hand, AI will simply conclude that tomorrow Jake will take out another $400, refraining from any moral judgment.

By the same token, what Twitter's algorithms infer from your followers and Twitter activity is that you may like X or Y content, without understanding whether you are right or left wing, smart or dumb, curious or narrow-minded. Importantly, especially if you have been binge-watching *Cobra Kai*, *Tiger King*, and *Bling Empire*, Netflix's algorithms won't assume that you are shallow and unintellectual. This is something only humans do, as it is humans' preoccupation to explain that this was only a result of lockdown boredom or an ironic and morbid curiosity with low and popular culture—at least that is my excuse.

A world of greater understanding would help us avoid many bad life choices. Just as it is easier to avoid wasting time in airports when you know there is a flight delay, or waiting to meet a friend when you know he's typically late, it would be much easier to minimize the risk of bad critical life choices when we are able to not just predict but also understand the likely outcome of these choices. For instance, people end up in unflattering, hideous, and unrewarding careers because someone told them it was a good idea, perhaps their aunt, uncle, mother, or cousin: "Go be a lawyer like your Uncle Tom." Imagine, instead, if they could actually understand that a different choice of job or career would see them thriving, performing well, succeeding.

TEST YOURSELF

Are You Curious?

How much (or how little) curiosity do you display in your everyday life?

- I rarely spend time reading books.
- I prefer to read headlines than entire stories.
- The good news about social media is that you can keep up with the news without studying things in detail.
- I was a lot more curious in my younger days.
- I have little time to ask why.
- So long as things work, I am not terribly interested in the details.
- There's no point in studying stuff when you can get any answer you want online.
- I waste little time daydreaming about stuff.
- I have no desire to understand why people have the wrong opinions.
- My friends and I think alike.

Add one point for every statement you agree with; then add up your points.

0–3: You may be living in a different age, or universe, or perhaps you are an exceptionally curious human. Despite technological distractions, you find time to exercise your mental muscles and feed your hungry mind.

4–6: You are average and within a range that can easily be flexed into the direction of more or less curiosity. Like most people, you should be careful that technological tools and other digital efficiencies don't hijack your desire to learn.

7–10: You are the perfect citizen of the AI age. Remember that machines may have the answer to any question, but this is only useful if you are actually able to ask any meaningful or relevant questions and critically evaluate the answers you get. It's time to unleash your ability to learn.

Chapter 8

How to Be Human

Toward a more humane AI age

You may not control all the events that happen to you, but you can
decide not to be reduced by them.

—Maya Angelou

The German explorer Alexander von Humboldt wrote that the aim
of existence is a "distillation of the widest possible experience of life
into wisdom."[1]

As this book has attempted to highlight, an honest self-
assessment of humanity in this early phase of the AI age suggests
that we are probably far from applying von Humboldt's principle.
Instead, it would be more accurate to say that the aim of our existence these days is to increase the wisdom of machines.

Whatever you think of humanity, it is not one click away. On
the contrary, we may be departing from it, one click at a time. Staring at a phone or computer screen all day, often merely to look at
our faces and at times to allocate ten-second attentional intervals

to a flurry of repetitive activities, apparently isn't distilling a wide range of experiences into wisdom. AI, on the other hand, seems to be following von Humboldt's precepts more closely than we do. AI today can experience humanity in all its dimensions—the good, the bad, the boring, and the useless.

Throughout history, humans have generally refrained from questioning their own utility, a bit like the turkey canceling Thanksgiving dinner. By the same token, we are probably not the most objective in judging the importance of our own species, not least since we aren't even objective to begin with. If we want to determine whether humans are likely to be automated or evolve as a species, humans may not be the most objective source of an answer. As playwright William Inge noted, "It is useless for the sheep to pass resolutions in favor of vegetarianism, while the wolf remains of a different opinion."

At the same time, the world is the product of human progress and inventiveness. Everything we've built, including AI, is the result of our creativity, talent, and ingenuity. This holds true even when many of our own inventions have resulted in making people less useful. Most famously, the industrial revolution led to systematic technological unemployment, a type of structural unemployment that is commonly attributed to technological innovations.[2] So, millions of artisan weavers became productively irrelevant, not to mention poor, when mechanized looms were introduced in order to produce *more with less*, the universal goal of technology, and the "less" part of this slogan tends to refer to humans.[3]

The good news is that there are no obvious signs that this is about to happen again any time soon. Just as in previous technological revolutions, AI is making certain tasks redundant while creating many new ones, keeping humans rather busy, provided they have the skills and motivation to partake in the new activities AI has created (e.g., store managers becoming e-commerce supervisors, e-commerce supervisors becoming cybersecurity analysts, and cybersecurity analysts becoming AI ethicists). The bad news, however, is that, irrespective of whether we can remain professionally active or useful, the AI age seems to bring out some of the worst in us. Rather than raising the psychological standards of humanity, it has lowered them, turning us into a perfectly dull and primitive version of ourselves.

This, I think, is the biggest tragedy of the AI age: while we were worrying about the automation of jobs and work, we have managed to somehow automate ourselves and our lives, injecting a strong dose of sanitized monotony and standardization into what was once a relatively interesting and fun life, at least when we look back from our current vantage point. While we are helping AI to upgrade itself, we are steadily downgrading ourselves. We can blame the Covid-19 pandemic and lockdowns for this dilution in cultural creativity and human imagination, which, compared with the real tragedy of those who got sick or lost their jobs or lives, would be best described as a #firstworldproblem. Perhaps a more accurate interpretation is that the pandemic has merely reminded us of the important things we lost—*real* rather than digital connectedness, a variety of experiences, and the richness of analogue adventures—by confining us to a form of life that resembles the hybrid predecessor of the metaverse. Fortunately, we can still ask whether there are better ways to express our humanity and strive for a more human and humane way of life in the AI age.

Admittedly, there are many ways of being human, and culture plays a strong role in determining the notable behavioral and attitudinal differences among large groups of people, including the observable commonalities, generalizations, and patterns that make *history*, once defined as "one damn thing after another," a meaningful and captivating human story.[4] If you have traveled abroad—which, at the time of writing this, requires a fair amount of imagination—you will have noticed these differences as soon as you landed in a foreign airport. Time, like culture and geography, is a potent influence on social behaviors, providing the basic norms and rules we use to express our humanity. So, there was a particular way of being human in the Roman empire, the Middle Ages, the Renaissance, and the original industrial revolution, and so on. That said, *within* each age or era, there are still more *or* less humane ways to express our humanity, so even if AI is unleashing some of our least desirable tendencies, on an individual level we all have the power and ability to resist this.

Chasing Happiness

The dehumanizing and sterilizing effects of technological efficiency on our lives represent a powerful fuel for positive psychology, a spiritual movement aimed at reclaiming some of the lost joy and fulfillment of being human. Positive psychology postulates that the essence of our being should be to transcend ourselves, attain spiritual equilibrium, and optimize our lives for positive emotions, such as subjective well-being and happiness. "Be happy" is the defining mantra of our times, permeating much of the meaning and purpose many people hope to extract from work and life. This proposition is now so mainstream that it seems hard to believe that our obsession with happiness is a fairly recent phenomenon, first propelled by the consumer society: chicken soup for the soul, uplifting melodies for our self-esteem.

Although there is nothing wrong with chasing happiness, academic research has long indicated that we are not very good at it, particularly when we obsess about it. From an evolutionary standpoint, it makes sense for happiness to be a moving target. As Oscar Wilde noted, "In this world there are only two tragedies. One is not getting what one wants, and the other is getting it."[5] But in fact, this is not equally true for everyone. Some people are prewired for happiness, which is why biologically determined personality traits, such as emotional stability, extraversion, and agreeableness (what academic psychologists call "EQ"), are a better predictor of lifelong happiness than external life events or circumstances, including winning the lottery or getting married—or for some, divorced.[6] Of course, we are all capable of experiencing happiness, but the point is that there is huge variability in the degree to which people need it, want it, and seek it, so telling someone to be happier is like telling someone to be more extraverted, less agreeable, and so on—like asking them to change their personality.

Happiness can at times turn into a pretty selfish or narcissistic goal, and there is evidence that the more satisfied you are with your life, especially your personal accomplishments, the more narcissistic you are.[7] How pleased you are when you look at yourself in the mirror says more about your ego than about the actual person in the mirror, particularly their objective accomplishments. If it feels good to be great, that's because you feel that you are great to begin with.

Leaving narcissism aside, there is no shortage of examples of humans optimizing their lives for happiness without contributing to any form of progress and in many ways obstructing it for everyone else. If your happiness is other people's misery, and your misery is other people's happiness, it is clear that something isn't working well. Contrary to popular belief, dissatisfaction, anger, and unhappiness are not only powerful but also productive human forces, so we shouldn't default to medicating ourselves against them. For every happiness pill or drug we consume—including the self-enhancing feedback we extract from AI-fueled social media platforms—a potential drive or ambition is numbed.

Throughout much of the industrialized world, the pursuit of happiness has turned into a biochemical quest, especially in the United States, which accounts for around 40 percent of the global pharma market, with just 5 percent of the world's population.[8] Yet if history is the biography of great people, their accomplishments are the product of their discontent rather than their blissful mindset. The Declaration of Independence mentions the "pursuit" rather than attainment of happiness, and it is not uncommon for the individual who is greatly focused on pursuing happiness to end up rather unhappy, even if in the process they end up attaining valuable things.

Any innovation or milestone in the evolution of human civilization is the result of people who were profoundly *un*happy with the status quo, which is precisely why they acted upon their grumpy discontent to replace established norms, products, and ideas with better ones. The ancient Greeks created one of the most advanced and sophisticated civilizations in human history, yet they were "passionate, unhappy, and at war with themselves."[9] This is also true for all ambitious modern corporations, whether Tesla, Goldman Sachs, or Facebook, which, admittedly, may not strike us as the most obvious progress-enhancing innovations in human history. After all, for the greatest portion of our evolution, we have somehow managed to exist without telling the world what our dog had for breakfast or discovering what our school classmates look like two decades later, which, thanks to *schadenfreude*, can at times make us feel happier about ourselves, at least temporarily. And yet, to many of the

2.3 billion people spending an average of forty minutes of their daily lives on Facebook, life would be far less enjoyable without it. Even after accounting for all the deep fakes, Belarussian hackers, and Russian bots, we are talking about 40 percent of the human race.

Importantly, if you are reluctant to accept that these examples indicate much progress, then remember that your unhappiness is as important a driver or incentive to change things. As Maya Angelou famously stated, "If you don't like something, change it." Equally, most of the groundbreaking innovations in the arts, science, and technology may have never been produced if their creators had been happy or interested simply in having a good time, feeling good, or having fun, as opposed to working hard in the pursuit of other people's happiness and collective progress.

Chasing happiness would be far more acceptable if we approached this from an *other*- rather than a *self*-perspective. Unlike in the individualistic West, people in collectivistic Eastern cultures are more inclined to harness happiness in others than themselves. So, what if the ultimate expression of our humanity were to live our lives trying to make *other* people happy or at least reduce their level of misery? What if, as Gandhi noted, "the only way to find yourself is to lose yourself in the interest of others"? Seen from this perspective, our ability to contribute to other people's happiness should be seen as more meaningful than our own personal happiness, and perhaps even a precondition for it.

Humanity Is a Work in Progress

Where will humans go? We should not decide on the basis of where AI can take us, for we are still in the driver's seat, even if it often doesn't feel that way. Driverless cars are not here yet, and hopefully we are not waiting for driverless humans anytime soon. So long as we still have a soul, and the ability to listen to it, we can avoid losing out, giving up, and fading into digital oblivion forever. We are more than a data-emission vessel and should cherish the ability to love, cry, smell, and smile, as these are still inherently and exclusively human activities. Pet lovers may disagree, but they are probably just projecting, though admittedly AI is still not as smart or obedient as my dog.

Even as AI and technology attempt to optimize our lives, and our *wants* and *wonts* are fully mediated by 0s and 1s, we are still largely busy with the same set of ancient activities: we learn, we talk, we work, we love, we hate, we bond, and in the middle of it all, we alternate between some certainty and a far larger dose of confusion, such as trying to understand why we are here to begin with. Perhaps *not* to get liked or retweeted?

At any given point in time and history, from Paleolithic cavemen to Instagram influencers, humans have always needed to get along with one another, get ahead of one another, and find meaning and purpose in their main life activities (e.g., religion, science, hobbies, work, or relationships). To the degree that we can leverage AI, or indeed any other tool we invent, to improve our ability to fulfill these needs—and assuming this makes us better in the process—we can feel optimistic about our own evolution.

Ultimately, the most important request is that humans try to better themselves, that is, find a better way of being human and improve humanity by making themselves better. As philosopher Simon Blackburn concludes in his brilliant short introduction to ethics, *Being Good*: "If we are careful, and mature, and imaginative, and fair, and nice, and lucky, the moral mirror in which we gaze at ourselves may not show us saints. But it need not show us monsters, either."[10] We all have the power to become a better version of ourselves, but that requires not just willpower but a sense of direction. If you are running very fast in the wrong direction, you will only end up further away from where you need to be, fooled by the illusion of progress and activity, and you will only manage to get lost quicker. As Lewis Carroll's cat tells Alice, "If you don't care much about where you want to go, then it really doesn't matter which way you go."[11]

We can do good, just as we are capable of doing bad even when we have moderate levels of intelligence and maturity. Adam Smith, the famous champion of free-market capitalism, is often evoked in the context of the "invisible hand" that allegedly manages to turn greedy and selfish for-profit motives into a benevolent order of things. Smith never really used that term, and a complete understanding of his theory and ideas ought to include his acknowledgment of the key importance of caring, empathy, and selfless drives, or as he notes: "How selfish soever man may be supposed, there are evidently some principles in his nature, which interest him in the fortune of others, and render their happiness necessary to him, though he derives nothing from it except the pleasure of seeing it. Of this kind is pity or compassion, the emotion which we feel for the misery of others, when we either see it, or are made to conceive it in a very lively manner."[12]

Simply put, groups are better off when their individuals behave in unselfish ways, because they can collectively benefit from one another's behaviors. Team performance is only possible if people focus less on their own individual goals and more on the goals of the team, and one person's sacrifices are another person's privileges. But this counterintuitive dynamic is not a natural state of affairs in groups, organizations, or societies. They need leadership to articulate or at least manage the tension between individual greed and collective productivity. The welfare of groups and societies cannot exist if its members are purely individualistic and selfish creatures. Take recycling: it is far less taxing and time-consuming to avoid it, but if everybody does so, then the environment collapses. That said, a few individuals can behave like leeches or parasites by refraining from recycling, so long as everyone else does. Or paying taxes: when Donald Trump bragged about being too smart to pay taxes, what he assumed was many people in the country do not care enough about this as much as we would expect, which explains why he is the most widely voted presidential candidate in US history (137.2 million votes across two elections).

As I write this, the world is witnessing another act of counterproductive greed around vaccine inequalities, with rich countries hoarding Covid vaccines beyond their needs while poor countries keep collapsing from the virus. Japan, Canada, and Australia have fewer than 1 percent of the world's Covid-19 cases but secured more vaccines than Latin America and the Caribbean, which are home to almost 20 percent of cases. You may think this is purely the result of meritocracy and therefore fairness in action, yet even rich countries lose when poor countries don't get vaccinated. So, it is not just that sheer altruism is missing but that greed is self-destructive.

But there is a catch: the kinder and more caring a group or system is, the bigger the incentive to not be kind. Not only is there less need for your kindness, because everyone is nice, but you also have more to gain by being a selfish leech and getting fat on other people's kindness—so long as you are part of a small minority of leeches. As soon as the leeches outnumber the Samaritans, competition *within* the system will weaken the system's ability to compete with other groups.

The most effective forms of group behavior manage to find perfect equilibrium between individual ambition and considerate empathy and prosocial orientation toward others. Such groups understand that individual success can't come at the expense of group well-being, and that group well-being is not possible without unleashing individual potential. Just as Adam Smith has an unfair reputation for promoting ruthless and cut-throat capitalism, while in reality he put kindness and empathy at the center of highly functioning societies, Darwin is usually bastardized as a proponent of dog-eats-dog Darwinism, when he really saw altruism and ethics as fundamental virtues for adaptive group survival and competitiveness: "Although a high standard of morality gives but a slight or no advantage to each individual man and his children over the other men of the same tribe . . . an advancement in the standard of morality will certainly give an immense advantage to one tribe over another."[13]

In the early days of the AI age, our main concern was that social media and excessive digital self-focus could make us selfish and antisocial, which, as shown in this book, studies have by now corroborated. But it is certainly too soon to conclude that the AI age causes systematic cultural shifts in selfishness. If kindness were a given, a default reaction, then we wouldn't be spending so much time lamenting that we don't have enough of it, or encouraging more people to be kind. Still, some people are kinder than others, and some cultures are kinder than others. And the main point is that kinder societies are better off, at least when you optimize for the majority of people rather than those who are at the top, especially when it isn't kindness that helps people get there. Therein lies the problem: when we try to make the world kind, then unkind people can take advantage of others and climb the hierarchy through sheer exploitation of people's kindness. And when we try to make the world unkind, then we lament the scarcity of kindness while the most brutally unkind crawl up the ladder with parasitic toxicity.

How, then, can we nurture moral kindness? We can't even get enough people to recycle or to genuinely care about the planet, which incidentally leads to their own self-destruction. Could AI help? Google spent the first ten years of its existence reminding itself (and us) that it must not be evil and a great proportion of the next ten years settling lawsuits, cleaning up its reputation, and firing its AI ethics researchers.[14] At least, today, worrying about the potential consequences of autonomous AI systems capable of acquiring immoral or unethical motives seems like the perfect excuse for not questioning our own moral standards.

In the quest to get better or become a better version of ourselves, we must demand that AI plays a bigger, more impactful role than it has played so far. If we can use technology to increase self-awareness and provide us with better self-understanding— including the things we may not like so much about ourselves— and highlight a gap between the person we are and the person we'd like to be, then there is clearly an opportunity to turn AI into a self-improvement tool and partner. Importantly, we don't even need AI to make huge advances in order to achieve this. It is possible, feasible, and straightforward to manage this even today.

There is no universal definition of progress, particularly at the individual human level. Someone may want to run a marathon, another may want to write a novel, and another may want to establish an empire. There are many different ways to conceptualize success, and each approach or interpretation has been encapsulated in one psychological model or another, particularly taxonomies of human values, which can group them into different buckets, such as status, freedom, joy, security, and so on. But one thing seems clear nonetheless: namely, that change is always part of the equation. You don't get better by staying the same or keeping things as they are.

Even for those who, like you and I, are privileged to belong to the educated proportion of the labor market known as knowledge workers, and who are less threatened by job automation and more likely to enjoy the benefits of improved working conditions and the evolution of work, our daily experience of work seems not too different from that of an alienated factory worker or the people in the industrial revolution. Objectively speaking we are better off, yet subjectively and vis-à-vis our rising and unrealistic expectations, the pressure to fulfill our dreams via our real work experiences may be too much of a psychological or spiritual burden. No wonder most people are disappointed, disengaged, and looking for another job, career, or life.

We need more magical experiences, and they will not happen online. Although we may not realize it, which speaks to the shrewd nudging tactics of Big Tech, there's a difference between optimizing our lives and optimizing the performance of the algorithms. There is also a difference between making our life easier and making the world a better place. In a rather counterintuitive fashion, our ability to increase the efficiencies of life, which today is largely achieved through algorithmic optimization, may be killing the very drive to deploy our human ingenuity to change the world for the better. The more satisfied we are with our everyday conveniences and efficiencies, the less likely we are to inject radical changes and innovations in the world.

Our Opportunity to Upgrade Humanity

Our opportunity is clear: to leverage AI to upgrade rather than downgrade or dilute our humanity. We have a significant chance to evolve as a species if we can capitalize on the AI revolution to make work more meaningful, unlock our potential, boost our understanding of ourselves and others, and create a less biased, more rational, and meaningful world. But there's a catch: we will only be able to achieve this if we can first acknowledge the potential risks that AI amplifies our less desirable, more destructive, and counterproductive tendencies. We have to use AI to unlock or harness our potential rather than to undermine or handicap our own existence. Quite clearly, there are important lessons from the rise of AI on how different dimensions of our humanity are not just being expressed and exposed but also reshaped.

What major expressions of our humanity will AI evoke or elicit? Will AI ultimately make or break us? Will it confer a significant economic, social, and cultural advantage to those societies that learn to master it? While we don't know how AI will ultimately change us, we can assume that some degree of change has already taken place—some good, some bad, and perhaps most of it undetectable. As author Margaret Visser remarked, "The extent to which we take everyday objects for granted is the precise extent to which they govern and inform our lives."[15]

Historian Melvin Kranzberg once noted that technology is neither good, nor bad, nor neutral. In fact, the only way for technology to *not* affect our humanity is for no one to use it, which is exactly the opposite of where we are today.[16] It has also been said that humans are prone to overestimating the short-term impact of technology but underestimating its long-term impact.[17]

Our children and grandchildren will probably learn about our predigital way of being through all our stories accessible via YouTube or its future equivalent. As in the pre-Sumerian civilizations, where legends and stories were passed orally from generation to generation through songs and plays, they shall hear of our adventurous dial-up explorations into a vastly unpopulated web, made of rustic digital landscapes devoid of Zoom fatigue, cat filters, and deep fakes, in which nobody knew we were dogs.

Humanity is a work in progress. This can only be a good thing. The alternative would mean that we have already reached the pinnacle of our evolution. Luckily, we haven't yet peaked, and there are clear signs that mankind is still developing and, yes, getting better.[18] This is not about colonizing Mars, building driverless cars, mastering quantum computing, or printing our perfect spouse in 4D, but about upgrading *ourselves*: creating a more adaptable, improved, and future-ready version of us.

You are the only person in the world whom you are guaranteed to influence, though it is not always easy. Even when others are able to influence you, it is only because you let them. Likewise when you are able to change others, it is only because you are able to influence your own behavior first. The wish may be overly optimistic and naive, at least if we accept the historical fact that humans have generally had very little incentive to out-behave their predecessors on the basis of evolutionary, let alone moral grounds. And yet, if we create the conditions that incentivize people to raise their own bar—looking for opportunities to stretch their potential—then we may see human civilization make progress, bottom-up, in an organic and incremental fashion.

Perhaps in the future we will take pleasure in escaping the web of predictions from machines and find enchanting moments of serendipity symbolizing that our creativity, ingenuity, and imagination are still intact. We may find ourselves in those magical white spaces we invent and produce, far from the reach of the algorithms, expanding our existence into the almost forgotten capacity for surprise, or at least self-surprise. An existence in which we deliberately trick AI to escape the boring and repetitive syntax of our existence and rewrite the grammar of life according to feelings, ideas, and acts that are of little interest and value to machines but deeply relevant to us. A life in which machines don't downgrade our intelligence and AI doesn't turn us into machines.

We can reclaim some of the rich variety of human experience and rediscover the balance between a somewhat algorithmic and efficient existence, on the one hand, and a fun, unpredictable, magical experience of life, on the other. We should not be the last generation to exist outside the matrix or the first to have its soul swallowed by machine learning. Instead, we should try to flourish. We can allow machines to continue learning, without ceasing to learn ourselves. We can outsmart AI, simply by not diluting our own intellectual capabilities or outsourcing our lives to computer algorithms. There is no point in automating any parts of our lives, let alone our entire existence, unless we have a plan for reinvesting that freedom and with it our capabilities and worthwhile activities. We have the power to be the main beneficiaries of AI and any other technology we invent and deploy, as opposed to becoming its product. Let us hope we can also find the will.

While the solution to our problems is far from clear, let alone simple, it will surely have to incorporate a combination of kindness, wisdom, and ingenuity. As Noam Chomsky notes:

> We're human beings, we're not automatons. You work at your job but you don't stop being a human being. Being a human

being means benefiting from rich cultural traditions—not just your own traditions but many others—and becoming not just skilled but also wise. Somebody who can think—think creatively, think independently, explore, inquire—and contribute to society. If you don't have that, you might as well be replaced by a robot. I think that simply can't be ignored if we want to have a society that's worth living in.[19]

The future starts today. The work starts now.

Notes

Introduction

1. Cami Rosso, "20 Great Quotes on Artificial Intelligence," *Psychology Today*, May 18, 2018, https://www.psychologytoday.com/us/blog/the-future-brain/201805/20-great-quotes-artificial-intelligence.

2. T. Chamorro-Premuzic and R. Akhtar, "Should Companies Use AI to Assess Job Candidates?," *Harvard Business Review*, May 2019, https://hbr.org/2019/05/should-companies-use-ai-to-assess-job-candidates.

3. F. Leutner, R. Akhtar, and T. Chamorro-Premuzic, *The Future of Recruitment: Using the new science of talent analytics to get your hiring right* (Bingley, UK: Emerald Group Publishing, 2022); A. Remonda, E. Veas, and G. Luzhnica, "Comparing Driving Behavior of Humans and Autonomous Driving in a Professional Racing Simulator," *PLoS One* 16 (2021); A. S. Ahuja, "The Impact of Artificial Intelligence in Medicine on the Future Role of the Physician," *PeerJ* 7 (2019); N. Dhieb et al., "A Secure AI-Driven Architecture for Automated Insurance Systems: Fraud detection and risk measurement," *IEEE Access* 8 (2020): 58546–58558.

4. D. Markovits, *The Meritocracy Trap* (New York: Penguin Books, 2019).

5. M. Appel, C. Marker, and T. Gnambs, "Are Social Media Ruining Our Lives? A review of meta-analytic evidence," *Review of General Psychology* 24 (2020): 60–74.

Chapter 1

1. Tomas Chamorro-Premuzic, "Selfie Sticks Should Be Banned for Massaging Our Self-Obsession," *Guardian*, August 27, 2015, https://www.theguardian.com/media-network/2015/aug/27/selfie-stick-self-obsession-narcissism-technology.

2. W. Durant and A. Durant, *The Lessons of History* (New York: Simon & Schuster, 2010).

3. M. Kruppa, "Venture Capitalists Seek Big Returns with NFTs," *Financial Times*, May 13, 2022.

4. W. Saletan, "Why Won't They Listen?," *New York Times*, March 3, 2021, https://www.nytimes.com/2012/03/25/books/review/the-righteous-mind-by-jonathan-haidt.html.

5. M. J. Guitton, "Cybersecurity, Social Engineering, Artificial Intelligence, Technological Addictions: Societal challenges for the coming decade," *Computers in Human Behavior* 107 (2020): 106–307.

6. Statista, "Daily Social Media Usage Worldwide," June 21, 2022, https://www.statista.com/statistics/433871/daily-social-media-usage-worldwide/.

7. C. Shoard, "Back to the Future Day: What part II got right and wrong about 2015–an A-Z," *Guardian*, January 2, 2015, https://www.theguardian.com/film/filmblog/2015/jan/02/what-back-to-the-future-part-ii-got-right-and-wrong-about-2015-an-a-z.

8. M. Chafkin, *The Contrarian: Peter Thiel and Silicon Valley's pursuit of power* (New York: Penguin Press, 2021).

9. R. Giphart and M. van Vugt, *Mismatch: How our stone age brain deceives us every day (and what we can do about it)* (London: Robinson, 2018).

10. R. Hogan and T. Chamorro-Premuzic, "Personality and Career Success," in *APA Handbook of Personality and Social Psychology*, vol. 4 (Washington, DC: American Psychological Association, 2014), 619–638, doi:10.1037/14343-028.

11. S. C. Matz, R. Appel, and K. Kosinski, "Privacy in the Age of Psychological Targeting," *Current Opinion in Psychology* 31 (2020): 116–121.

12. D. T. Wegener and R. E. Petty, "The Naive Scientist Revisited: Naive theories and social judgment," *Social Cognitive and Affective Neuroscience* 16 (1998) 1–7.

13. T. Chamorro-Premuzic, "Humans Have the Power to Remain Socially and Emotionally Connected Even in Extreme Physical Isolation," *Forbes*, March 20, 2020, https://www.forbes.com/sites/tomaspremuzic/2020/03/20/even-in-physical-isolation-we-will-remain-socially-and-emotionally-connected/?sh=68ec879b458a.

14. T. Chamorro-Premuzic, "Thriving in the Age of Hybrid Work," *Harvard Business Review*, January 2021, https://hbr.org/2021/01/thriving-in-the-age-of-hybrid-work.

15. C. Hanson, "SXSW 2021: Looking through the lens of commerce," OMD, March 29, 2021, https://www.omd.com/thoughts/sxsw-2021-looking-through-the-lens-of-commerce/.

16. Y. N. Harari, *21 Lessons for the 21st Century* (New York: Random House, 2018).

17. V. Kazuo Ishiguro and Venki Ramakrishnan, "Imagining a New Humanity," *Financial Times*, May 25, 2021.

18. Ibid.

19. Ibid.

20. Ibid.

21. M. Kosinski, D. Stillwell, and T. Graepel, "Private Traits and Attributes Are Predictable from Digital Records of Human Behavior," *Proceedings of the National Academy of Science* 110 (2013): 5802–5805.

22. J. Vincent, "Twitter Is Bringing Its 'Read Before You Retweet' Prompt to All Users," *Verge*, September 25, 2020, https://www.theverge.com/2020/9/25/21455635/twitter-read-before-you-tweet-article-prompt-rolling-out-globally-soon.

23. Ibid.

24. P. Olson, "Facebook Closes $19 Billion WhatsApp Deal," *Forbes*, October 6, 2014, https://www.forbes.com/sites/parmyolson/2014/10/06/facebook-closes-19-billion-whatsapp-deal/?sh=737c1efe5c66; A. L. Deutsch, "WhatsApp: The best Meta purchase ever?," Investopedia, March 29, 2022, https://www.investopedia.com/articles/investing/032515/whatsapp-best-facebook-purchase-ever.asp.

25. "Fixing Economic Inequality: Lawrence Summers," Rotman School of Management, YouTube video, April 7, 2015, https://www.youtube.com/watch?v=wXMEoS7OsO0.

26. G. Deyan, "How Much Time Do People Spend on Social Media in 2022?," *techjury* (blog) June 2, 2022, https://techjury.net/blog/time-spent-on-social-media/#gref.

27. Y. Bikker, "How Netflix Uses Big Data to Build Mountains of Money," Medium, 2020, https://medium.com/swlh/how-netflix-uses-big-data-to-build-mountains-of-money-829364caefa7; B. Marr, "The Amazing Ways Spotify Uses Big Data, AI and Machine Learning to Drive Business Success," *Forbes*, October 30, 2017, https://www.forbes.com/sites/bernardmarr/2017/10/30/the-amazing-ways-spotify-uses-big-data-ai-and-machine-learning-to-drive-business-success/?sh=7bb1ce464bd2.

28. "Rethinking How We Value Data," *Economist*, February 27, 2020, https://www.economist.com/finance-and-economics/2020/02/27/rethinking-how-we-value-data.

29. O. Staley, "Tech Stocks Powered the S&P 500 Index to a Monster Performance in 2021," Quartz, October 29, 2021, https://qz.com/2108056/apple-amazon-microsoft-and-alphabet-drove-the-sp-500-in-2021/; SaaS Capital, "2021 Private SaaS Company Valuations," 2021, https://www.saas-capital.com/blog-posts/2021-private-saas-company-valuations/.

30. A. Agrawal, J. Gans, and A. Goldfarb, *Prediction Machines: The simple economics of artificial intelligence* (Boston: Harvard Business Review Press, 2018).

31. PricewaterhouseCoopers, "Sizing the Prize: What's the real value of AI for your business and how can you capitalize?," PwC, 2021, https://www.pwc.com/gx/en/issues/data-and-analytics/publications/artificial-intelligence-study.html#:~:text=AI could contribute up to,come from consumption-side effects.

32. M. Graham and J. Elias, "How Google's $150 Billion Advertising Business Works," CNBC, May 18, 2021, https://www.cnbc.com/2021/05/18/how-does-google-make-money-advertising-business-breakdown-.html.

33. M. Johnston, "How Facebook (Meta) Makes Money: Advertising, payments, and other fees," Investopedia, July 17, 2022, https://www.investopedia.com/ask/answers/120114/how-does-facebook-fb-make-money.asp.

34. G. Kuhn, "How Target Used Data Analytics to Predict Pregnancies," Drive Research, July 16, 2020, https://www.driveresearch.com/market-research-company-blog/how-target-used-data-analytics-to-predict-pregnancies/.

35. J. Naughton, "'The Goal Is to Automate Us': Welcome to the age of surveillance capitalism," *Guardian*, January 20, 2019, https://www.theguardian.com/technology/2019/jan/20/shoshana-zuboff-age-of-surveillance-capitalism-google-facebook.

36. C. Campbell, "How China Is Using Big Data to Create a Social Credit Score," *Time*, 2019, https://time.com/collection/davos-2019/5502592/china-social-credit-score/.

37. Mozilla Foundation, "Match.com—Privacy & Security Guide," March 15, 2021, https://foundation.mozilla.org/en/privacynotincluded/matchcom/.

38. S. Zuboff, *The Age of Surveillance Capitalism: The fight for a human future at the new frontier of power* (New York: PublicAffairs, 2019).

39. A. Levy, "Tech's Top Seven Companies added $3.4 Trillion in Value in 2020," CNBC, December 31, 2020, https://www.cnbc.com/2020/12/31/techs-top-seven-companies-added-3point4-trillion-in-value-in-2020.html

40. *Fortune*, Fortune Global 500: Amazon, 2021, https://fortune.com/company/amazon-com/global500/.

41. F. Ali, "How Ecommerce and Small Businesses Were Affected by COVID-19," *Digital Commerce 360*, February 19, 2021, https://www.digitalcommerce360.com/2021/02/19/ecommerce-during-coronavirus-pandemic-in-charts/.

42. *Economist*, "Who Owns the Web's Data?," *Economist*, October 22, 2020, https://www.economist.com/business/2020/10/22/who-owns-the-webs-data.

43. *Wall Street Journal*, "Five Tech Giants Just Keep Growing," May 1, 2021, https://www.wsj.com/articles/five-tech-giants-just-keep-growing-11619841644; D. Rabouin, "Big Tech's Share of the S&P 500 Reached Record Level in August," Axios, September 28, 2020, https://www.axios.com/2020/09/28/big-techs-share-of-the-sp-500-reached-record-level-in-august.

44. *Wall Street Journal*, "Five Tech Giants Just Keep Growing."

45. A. Munro, "State of Nature (Political Theory)," *Encyclopedia Britannica,* 2020, https://www.britannica.com/topic/state-of-nature-political-theory.

46. T. W. Hsu et al., "Social Media Users Produce More Affect That Supports Cultural Values, But Are More Influenced by Affect That Violates Cultural Values," *Journal of Personality and Social Psychology* 121 (2021): 969–983.

47. S. Pulman and Frank Rose, "The Art of Immersion: How the Digital Generation Is Remaking Hollywood, Madison Avenue, and the Way We Tell Stories," *International Journal of Advertising* 30 (2011): 151–153.

48. G. Miller, *The Mating Mind: How sexual choice shaped the evolution of human nature* (New York: Vintage, 2001).

49. W. Durant and A. Durant, *The Lessons of History* (New York: Simon & Schuster, 2010).

50. R. Giphart and M. van Vugt, *Mismatch*.

51. J. R. Speakman, "Evolutionary Perspectives on the Obesity Epidemic: Adaptive, maladaptive, and neutral viewpoints," *Annual Review of Nutrition* 33 (2013): 289–317.

52. C. Burke, "Distracted in the Office? Blame Evolution," *Guardian*, March 1, 2016, https://www.theguardian.com/careers/2016/mar/01/distracted-office-blame-evolution-workspace-design-focus.

Chapter 2

1. H. A. Simon, "Designing Organizations for an Information-Rich World," in *Computers, Communication, and the Public Interest,* Martin Greenberger, ed. (Baltimore: Johns Hopkins University Press, 1971), 37–72.

2. N. Pope, "The Economics of Attention: Style and substance in the age of information," *Technology and Culture* 48 (2007): 673–675.

3. Nishat Kazi, "The Identity Crisis of Libraries in the Attention Economy," *Library Philosophy and Practice* (2012) 684.

4. S. Giraldo-Luque, P. N. A. Afanador, and C. Fernández-Rovira, "The Struggle for Human Attention: Between the abuse of social media and digital wellbeing," *Healthcare* 8 (2020): 497.

5. S. Kemp, "Digital 2022: Time spent using connected tech continues to rise," DataReportal, 2022, https://datareportal.com/reports/digital-2022-time-spent-with-connected-tech.

6. N. Slatt, "Apple Now Sells More Watches Than the Entire Swiss Watch Industry," *Verge*, February 5, 2020, https://www.theverge.com/2020/2/5/21125565/apple-watch-sales-2019-swiss-watch-market-estimates-outsold; *Economist*, "How a New Age of Surveillance Is Changing Work," *Economist*, May 13, 2022, https://www.economist.com/leaders/2022/05/13/how-a-new-age-of-surveillance-is-changing-work.

7. P. Lewis, "'Our Minds Can Be Hijacked': The tech insiders who fear a smartphone dystopia," *Guardian*, October 5, 2017, https://www.theguardian.com/technology/2017/oct/05/smartphone-addiction-silicon-valley-dystopia.

8. J. Hari, *Stolen Focus: Why you can't pay attention—and how to think deeply again* (New York: Crown, 2022).

9. N. G. Carr, *The Shallows: What the internet is doing to our brains* (New York: W. W. Norton & Company, 2010).

10. N. Carr, "Author Nicholas Carr: The web shatters focus, rewires brains," *Wired*, May 24, 2010, https://www.wired.com/2010/05/ff-nicholas-carr/.

11. R. Kay, D. Benzimra, and J. Li, "Exploring Factors That Influence Technology-Based Distractions in Bring Your Own Device Classrooms," *Journal of Educational Computing Research* 55 (2017): 974–995.

12. B. A. Barton et al., "The Effects of Social Media Usage on Attention, Motivation, and Academic Performance," *Active Learning in Higher Education* 22 (2021): 11–22.

13. S. Feng et al., "The Internet and Facebook Usage on Academic Distraction of College Students," *Computers and Education* 134 (2019): 41–49.

14. S. Giraldo-Luque, P. N. A. Afanador, and C. Fernández-Rovira, "The Struggle for Human Attention."

15. L. Miarmi and K. G. DeBono, "The Impact of Distractions on Heuristic Processing: Internet advertisements and stereotype use," *Journal of Applied Social Psychology* 37 (2007): 539–548.

16. J. Carpenter et al., "The Impact of Actively Open-Minded Thinking on Social Media Communication," *Judgment and Decision Making* 13 (2018): 562–574.

17. S. Brooks, P. Longstreet, and C. Califf, "Social Media Induced Technostress and Its Impact on Internet Addiction: A distraction-conflict theory perspective," *AIS Transcripts on Human-Computer Interaction* 9 (2017): 99–122.

18. T. Mahalingham, J. Howell, and P. J. F. Clarke, "Attention Control Moderates the Relationship between Social Media Use and Psychological Distress," *Journal of Affective Disorders* 297 (2022): 536–541.

19. S. Kautiainen et al., "Use of Information and Communication Technology and Prevalence of Overweight and Obesity among Adolescents," *International Journal of Obesity* 29 (2005): 925–933.

20. S. Anderson, "In Defense of Distraction," *New York Magazine*, May 15, 2009, https://nymag.com/news/features/56793/index1.html.

21. L. Stone, "Category Archives: continuous partial attention," *Linda Stone* (blog) January 19, 2014, https://lindastone.net/category/attention/continuous-partial-attention/.

22. I. Koch et al., "Switching in the Cocktail Party: Exploring intentional control of auditory selective attention," *Journal of Experimental Psychology: Human Perception and Performance* 37 (2011): 1140–1147.

23. Y. Jeong, H. Jung, and J. Lee, "Cyberslacking or Smart Work: Smartphone usage log-analysis focused on app-switching behavior in work and leisure conditions," *International Journal of Human–Computer Interaction* 36 (2020): 15–30.

24. *Economist*, "Are Digital Distractions Harming Labour Productivity?," *Economist*, December 7, 2017, https://www.economist.com/finance-and-economics/2017/12/07/are-digital-distractions-harming-labour-productivity.

25. *Economist*, "Are Digital Distractions Harming Labour Productivity?;" B. Solis, "Our Digital Malaise: Distraction is costing us more than we think," *LSE Business Review* (blog) April 19, 2019, https://blogs.lse.ac.uk/businessreview/2019/04/19/our-digital-malaise-distraction-is-costing-us-more-than-we-think/.

26. Information Overload Research Group, accessed September, 2022, https://iorgforum.org/.

27. P. Juneja, "The Economic Effects of Digital Distractions," Management Study Guide, n.d., https://www.managementstudyguide.com/economic-effects-of-digital-distractions.htm.

28. P. Bialowolski et al., "Ill Health and Distraction at Work: Costs and drivers for productivity loss," *PLoS One* 15 (2020): e0230562.

29. B. A. Barton et al., "The Effects of Social Media Usage on Attention, Motivation, and Academic Performance."

30. *Economist*, "Are Digital Distractions Harming Labour Productivity?"

31. Y. Jeong, H. Jung, and J. Lee, "Cyberslacking or Smart Work"; A. Tandon et al., "Cyberloafing and Cyberslacking in the Workplace: Systematic literature review of past achievements and future promises," *Internet Research* 32 (2022): 55–89.

32. Avery Hartmans, "Major US Tech Firms Keep Pushing Back Their Return-to-Office deadlines. Maybe It's Time to Admit Defeat,"*Business Insider*, December 12, 2021, https://www.businessinsider.com/mandatory-return-to-office-why-2021-12.

33. J. Goldman, "6 Apps to Stop Your Smartphone Addiction," *Inc.*, October 21, 2015, https://www.inc.com/jeremy-goldman/6-apps-to-stop-your-smartphone-addiction.html.

34. B. Dean, "Netflix Subscriber and Growth Statistics: How many people watch Netflix in 2022?," Backlinko, October 27, 2021, https://backlinko.com/netflix-users.

35. A. Steele, "Apple, Spotify and the New Battle over Who Wins Podcasting," *Wall Street Journal*, April 23, 2021, https://www.wsj.com/articles/apple-spotify-and-the-new-battle-over-who-wins-podcasting-11619170206?mod=article_inline.

36. A. Alter, "Why Our Screens Make Us Less Happy," TED Talk, 2017, https://www.ted.com/talks/adam_alter_why_our_screens_make_us_less_happy#t-189373.

37. A. K. Pace, "Coming Full Circle: Digital distraction," *Computer Library* 23 (2003): 50–51.

38. R. Hogan and T. Chamorro-Premuzic, "Personality and the Laws of History," in *Wiley-Blackwell Handbook of Individual Differences* (Hoboken, NJ: Wiley-Blackwell, 2011), 491–511, doi:10.1002/9781444343120.ch18.

39. G. J. Robson, "The Threat of Comprehensive Overstimulation in Modern Societies," *Ethics and Information Technology* 19 (2017): 69–80.

40. R. H. Fazio and M. P. Zanna, "Direct Experience and Attitude-Behavior Consistency," *Advances in Experimental Social Psychology* 14 (1981): 161–202.

41. J. Fuhrman, "The Hidden Dangers of Fast and Processed Food," *American Journal of Lifestyle Medicine* 12 (2018): 375–381.

Chapter 3

1. A. C. Edmonson and T. Chamorro-Premuzic, "Today's Leaders Need Vulnerability, Not Bravado," hbr.org, October 19, 2020, https://hbr.org/2020/10/todays-leaders-need-vulnerability-not-bravado.

2. C. Sindermann, J. D. Elhai, and C. Montag, "Predicting Tendencies Towards the Disordered Use of Facebook's Social Media Platforms: On the role of personality, impulsivity, and social anxiety," *Psychiatry Research* 285 (2020); J. Anderer, "Hurry Up! Modern patience thresholds lower than ever before," SurveyFinds, September 3, 2019, https://www.studyfinds.org/hurry-up-modern-patience-thresholds-lower-than-ever-before-survey-finds/.

3. T. S. van Endert and P. N. C. Mohr, "Likes and Impulsivity: Investigating the relationship between actual smartphone use and delay discounting," *PLoS One* 15 (2020): e0241383.

4. H. Cash et al., "Internet Addiction: A brief summary of research and practice," *Current Psychiatry Reviews* 8 (2012): 292–298.

5. A. Shashkevich, "Meeting Online Has Become the Most Popular way U.S. Couples Connect, Stanford Sociologist Finds," *Stanford News*, August 21, 2019, https://news.stanford.edu/2019/08/21/online-dating-popular-way-u-s-couples-meet/.

6. Tinder, "About Tinder," 2022, https://tinder.com/en-GB/about-tinder.

7. E. Charlton, "What Is the Gig Economy and What's the Deal for Gig Workers?," World Economic Forum, May 26, 2021, https://www.weforum.org/agenda/2021/05/what-gig-economy-workers/.

8. S. Manchiraju, A. Sadachar, and J. L. Ridgway, "The Compulsive Online Shopping Scale (COSS): Development and validation using panel data," *International Journal of Mental Health and Addiction* 15, (2017): 209–223.

9. Manchiraju, Sadachar, and Ridgway, "The Complusive Online Shopping Scale (COSS)."

10. D. Boffey, "Abolish Internet Shopping in Belgium, Says Leader of Party in Coalition," *Guardian*, February 11, 2022, https://www.theguardian.com/world/2022/feb/11/abolish-internet-shopping-belgium-says-paul-magnette-socialist-leader-coalition.

11. B. Pietrykowski, "You Are What You Eat: The social economy of the slow food movement," *Review of Social Economy* 62, no. 3 (September 2004): 307–321, https://doi.org/10.1080/0034676042000253927.

12. Haoran, "TikTok: Using AI to take over the world," Digital Innovation and Transformation, HBS Digital Initiative, April 19, 2020, https://digital.hbs.edu/platform-digit/submission/tik-tok-using-ai-to-take-over-the-world/.

13. M. Rangaiah, "What Is TikTok and How Is AI Making It Tick?," *Analytic Steps* (blog) January 16, 2020 https://www.analyticssteps.com/blogs/how-artificial-intelligence-ai-making-tiktok-tick.

14. *Wall Street Journal*, "Inside TikTok's Algorithm: A WSJ video investigation," July 21, 2021, https://www.wsj.com/articles/tiktok-algorithm-video-investigation-11626877477.

15. J. S. B. T. Evans, "In Two Minds: Dual-process accounts of reasoning," *Trends in Cognitive Science* 7 (2003): 454–459.

16. J. Ludwig, F. K. Ahrens, and A. Achtziger, "Errors, Fast and Slow: An analysis of response times in probability judgments," *Thinking & Reasoning* 26 (2020), doi:10.1080/13546783.2020.1781691.

17. D. Choi et al., "Rumor Propagation Is Amplified by Echo Chambers in Social Media," *Scientific Reports* 10 (2020).

18. S. W. C. Nikkelen et al., "Media Use and ADHD-Related Behaviors in Children and Adolescents: A meta-analysis," *Developmental Psychology Journal* 50 (2014): 2228–2241.

19. B. Sparrow, J. Liu, and D. M. Wegner, "Google Effects on Memory: Cognitive consequences of having information at our fingertips," *Science* 333 (2011): 776–778.

20. R. B. Kaiser and D. V. Overfield, "Strengths, Strengths Overused, and Lopsided Leadership," *Consulting Psychology Journal* 63 (2011): 89–109.

21. R. Hogan and T. Chamorro-Premuzic, "Personality and the Laws of History," *Wiley-Blackwell Handbook of Individual Differences* (Hoboken, NJ: Wiley-Blackwell, 2011), 491–511, doi:10.1002/9781444343120.ch18.

22. J. Polivy and C. Peter Herman, "If at First You Don't Succeed: False hopes of self-change," *American Psychologist Journal* 57 (2002): 677–689.

23. R. Boat and S. B. Cooper, "Self-Control and Exercise: A review of the bi-directional relationship," *Brain Plasticity* 5 (2019): 97–104.

24. R. F. Baumeister and J. Tierney, *Willpower: Rediscovering our greatest strength* (New York: Penguin Books, 2011).

25. C. L. Guarana et al., "Sleep and Self-Control: A systematic review and meta-analysis," *Sleep Medicine Reviews* 59 (2021): 101514.

26. A. J. Sultan, J. Joireman, and D. E. Sprott, "Building Consumer Self-Control: The effect of self-control exercises on impulse buying urges," *Marketing Letters* 23 (2012): 61–72.

27. R. Boat and S. B. Cooper, "Self-Control and Exercise."

Chapter 4

1. D. Ariely, *Predictably Irrational: The hidden forces that shape our decisions* (New York: Harper, 2008).

2. L. F. Barrett, *Seven and a Half Lessons about the Brain* (New York: Mariner Books, 2020).

3. D. Ariely, *Predictably Irrational*.

4. E. Pronin, D. Y. Lin, and L. Ross, "The Bias Blind Spot: Perceptions of bias in self versus others," *Personal and Social Psychology Bulletin* 28 (2002): 369–381.

5. W. Durant, *The Story of Philosophy: The lives and opinions of the greater philosophers* (New York: Pocket Books, 1991).

6. M. Hewstone, "The 'Ultimate Attribution Error'? A review of the literature on intergroup causal attribution," *European Journal of Social Psychology* 20 (1990): 311–335.

7. T. Sharot, "The Optimism Bias," *Current Biology* 21 (2011): R941–R945.

8. P. S. Forscher et al., "A Meta-Analysis of Procedures to Change Implicit Measures," *Journal of Personal and Social Psychology* 117 (2019): 522–559.

9. T. Beer, "All the Times Trump Compared Covid-19 to the Flu, Even After He Knew Covid-19 Was Far More Deadly," *Forbes*, September 10, 2020, https://www.forbes.com/sites/tommybeer/2020/09/10/all-the-times-trump-compared-covid-19-to-the-flu-even-after-he-knew-covid-19-was-far-more-deadly/?sh=109ffaa3f9d2; C. Hutcherson et al., "The Pandemic Fallacy: Inaccuracy of social scientists' and lay judgments about COVID-19's societal consequences in America," *PsyArXiv* (2021), doi:10.31234/OSF.IO/G8F9S; J. Demsas, "Why So Many COVID Predictions Were Wrong," *Atlantic*, April 2022, https://www.theatlantic.com/ideas/archive/2022/04/pandemic-failed-economic-forecasting/629498/.

10. K. Dou et al., "Engaging in Prosocial Behavior Explains How High Self-Control Relates to More Life Satisfaction: Evidence from three Chinese samples," *PLoS One* 14 (2019): e0223169; J. Passmore and L. Oades, "Positive Psychology Techniques: Random acts of kindness and consistent acts of kindness and empathy," *Coaching Psychologist* 11 (2015): 90–92.

11. K. Atkinson, T. Bench-Capon, and D. Bollegala, "Explanation in AI and Law: Past, present and future," *Journal of Artificial Intelligence* 289 (2020): 103387.

12. E. Tay Hunt, "Microsoft's AI Chatbot, Gets a Crash Course in Racism from Twitter," *Guardian*, March 24, 2016, https://www.theguardian.com/technology/2016/mar/24/tay-microsofts-ai-chatbot-gets-a-crash-course-in-racism-from-twitter.

13. J. Dastin, "Amazon Scraps Secret AI Recruiting Tool That Showed Bias Against Women," Reuters, October 10, 2018, https://www.reuters.com/article/us-amazon-com-jobs-automation-insight-idUSKCN1MK08G.

14. E. Pronin, D. Y. Lin, and L. Ross, "The Bias Blind Spot".

15. A. N. Kluger and A. DeNisi, "The Effects of Feedback Interventions on Performance: A historical review, a meta-analysis, and a preliminary feedback intervention theory," *Psychological Bulletin* 119 (1996): 254284.

16. F. Morse, "Facebook Dislike Button: A short history," BBC News, September 16, 2015, https://www.bbc.co.uk/news/newsbeat-34269663.

17. K. Scott, *Radical Candor: Be a kick-ass boss without losing your humanity* (New York: St. Martin's Press, 2019).

18. Charles Stangor, Rajiv Jhangiani and Hammond Tarry, "The Social Self: The role of the social situation," in *Principles of Social Psychology,* 1st International H5P ed. (1996), 1–36.

19. L. Uziel, "Rethinking Social Desirability Scales: From impression management to interpersonally oriented self-control," *Perspectives on Psychological Science* 5 (2010): 243–262.

20. R. Hogan, T. Chamorro-Premuzic, and R. B. Kaiser, "Employability and Career Success: Bridging the gap between theory and reality," *Industrial and Organizational Psychology* 6 (2013): 3–16.

21. E. Dhawan, *Digital Body Language: How to build trust and connection, no matter the distance* (New York: St Martin's Press, 2021).

22. P. Tajalli, "AI Ethics and the Banality of Evil," *Ethics and Information Technology* 23 (2021): 447–454.

23. B. Christian, *The Alignment Problem: Machine learning and human values* (New York: W. W. Norton & Company, 2015).

24. J. Gans, "AI and the Paperclip Problem," Vox, CEPR Policy Portal, June 10, 2018, https://voxeu.org/article/ai-and-paperclip-problem.

25. "SNP Genotyping," Wikipedia, 2022, https://en.wikipedia.org/wiki/SNP_genotyping.

26. V. Eubanks, *Automating Inequality* (New York: St. Martin's Press, 2018).

27. J. Drescher, "Out of DSM: Depathologizing Homosexuality," *Behavioral Science* 5 (2015): 565.

28. P. Costa, *The Edge of Democracy* (Netflix, 2019).

Chapter 5

1. M. Bergmann, "The Legend of Narcissus," *American Imago* 41 (1984): 389–411.

2. C. J. Carpenter, "Narcissism on Facebook: Self-promotional and anti-social behavior," *Personality and Individual Differences* 52 (2012): 482–486.

3. R. L. Kauten et al., "Purging My Friends List. Good Luck Making the Cut: Perceptions of narcissism on Facebook," *Computers in Human Behavior* 51 (2015): 244–254.

4. R. Chandra, "Is Facebook Making Us Narcissistic?," *Psychology Today*, February 5, 2018, https://www.psychologytoday.com/us/blog/the-pacific-heart/201802/is-facebook-making-us-narcissistic.

5. E. Grijalva and L. Zhang, "Narcissism and Self-Insight: A review and meta-analysis of narcissists' self-enhancement tendencies," *Personality and Social Psychology Bulletin* 42 (2016): 3–24.

6. E. Grijalva and L. Zhang, "Narcissism and Self-Insight."

7. I. Lunden and T. Hatmaker, "Twitter Accepts Elon Musk's $44B Acquisition Offer," TechCrunch, April 25, 2022, https://techcrunch.com/2022/04/25/twitter-accepts-elon-musks-43b-acquisition-offer/.

8. B. Gentile et al., "The Effect of Social Networking Websites on Positive Self-Views: An experimental investigation," *Computers in Human Behavior* 28 (2012): 1929–1933.

9. A. Rijsenbilt and H. Commandeur, "Narcissus Enters the Courtroom: CEO narcissism and fraud," *Journal of Business Ethics* 117 (2013): 413–429.

10. S. M. Bergman et al., "Millennials, Narcissism, and Social Networking: What narcissists do on social networking sites and why," *Personality and Individual Differences* 50 (2011): 706–711.

11. K. Nash, A. Johansson, and K. Yogeeswaran, "Social Media Approval Reduces Emotional Arousal for People High in Narcissism: Electrophysiological evidence," *Frontiers in Human Neuroscience* 13 (2019): 292.

12. A. Rijsenbilt and H. Commandeur, "Narcissus Enters the Courtroom."

13. O. Güell, "Social Media Addiction: Rise of selfie deaths leads experts to talk about a public health problem," *EL PAÍS*, October 29, 2021, https://english.elpais.com/usa/2021-10-29/rise-of-selfie-deaths-leads-experts-to-talk-about-a-public-health-problem.html.

14. B. Lindström et al., "A Computational Reward Learning Account of Social Media Engagement," *Nature Communications* 12 (2021).

15. A. K. Pace, "Coming Full Circle: Digital distraction," *Computer Library* 23 (2003): 50–51.

16. T. F. Heatherton and C. Wyland, "Why Do People Have Self-Esteem?," *Psychological Inquiry* 14 (2003): 38–41.

17. N. Chomsky, *Requiem for the American Dream: The 10 principles of concentration of wealth and power* (New York: Seven Stories Press, 2017).

18. L. Elliott, "World's 26 Richest People Own as Much as Poorest 50%, Says Oxfam," *Guardian*, January 21, 2019, https://www.theguardian.com/business/2019/jan/21/world-26-richest-people-own-as-much-as-poorest-50-per-cent-oxfam-report.

19. B. Kellerman and T. L. Pittinsky, *Leaders Who Lust: Power, money, sex, success, legitimacy, legacy* (New York: Cambridge University Press, 2020).

20. D. T. Hsu and J. M. Jarcho, "Next Up for Psychiatry: Rejection sensitivity and the social brain," *Neuropsychopharmacology* 46 (2021): 239–240.

21. L. D. Rosen et al., "Is Facebook Creating 'iDisorders'? The link between clinical symptoms of psychiatric disorders and technology use, attitudes and anxiety," *Computers in Human Behavior* 29 (2013): 1243–1254.

22. T. Chamorro-Premuzic, "Is Authenticity at Work Overrated?," *Fast Company*, November 12, 2018, https://www.fastcompany.com/90256978/is-authenticity-overrated.

23. T. Chamorro-Premuzic, "4 Pieces of Career Advice It's Okay to Ignore," Ascend, hbr.org, October 15, 2020, https://hbr.org/2020/10/4-pieces-of-career-advice-its-okay-to-ignore.

24. R. Hogan, T. Chamorro-Premuzic, and R. B. Kaiser, "Employability and Career Success: Bridging the gap between theory and reality," *Industrial and Organizational Psychology* 6 (2013): 3–16.

25. E. Goffman, *The Presentation of Everyday Life* (New York: Doubleday, 1959).

26. J. Pfeffer, *Leadership BS: Fixing workplaces and careers one truth at a time* (New York: Harper Business, 2015).

27. R. Lambiotte and M. Kosinski, "Tracking the Digital Footprints of Personality," *Proceedings of the IEEE* 102 (2014): 1934–1939.

28. T. Chamorro-Premuzic, *The Talent Delusion: Why data, not intuition, is the key to unlocking human potential* (London: Piatkus, 2017).

29. C. R. Colvin, J. Block, and D. C. Funder, "Overly Positive Self-Evaluations and Personality: Negative implications for mental health," *Journal of Personality and Social Psychology* 68 (1995): 1152–1162.

30. T. Chamorro-Premuzic, *Confidence: How much you really need and how to get it* (New York: Plume, 2014).

31. W. Durant and A. Durant, *The Lessons of History* (New York: Simon & Schuster, 2010).

32. T. Chamorro-Premuzic, *Why Do So Many Incompetent Men Become Leaders? (and How to Fix It)* (Boston: Harvard Business Review Press, 2019).

33. A. Edmondson and T. Chamorro-Premuzic, "Today's Leaders Need Vulnerability, Not Bravado," *Harvard Business Review*, October 19, 2020, https://www.hbs.edu/faculty/Pages/item.aspx?num=59153.

Chapter 6

1. M. J. Rosenfeld, R. J. Thomas, and S. Hausen, "Disintermediating Your Friends: How online dating in the United States displaces other ways of meeting," *Proceedings of the National Academy of Sciences of the United States of America* 116 (2019): 17753–1758.

2. Centers for Disease Control and Prevention, "Road Traffic Injuries and Deaths—A Global Problem," 2020, https://www.cdc.gov/injury/features/global-road-safety/index.html.

3. M. C. Marchese and P. M. Muchinsky, "The Validity of the Employment Interview: A meta-analysis," *International Journal of Selection and Assessment* 1 (1993): 18–26.

4. B. D. Spencer, "Estimating the Accuracy of Jury Verdicts," *Journal of Empirical Legal Studies* 4 (2007): 305–329.

5. A. R. McConnell et al., "The Simple Life: On the benefits of low self-complexity," *Personality and Social Psychology Bulletin* 35 (2009): 823–835.

6. S. Harris, *Free Will* (New York: Free Press, 2012).

7. A. Grant, "Feeling Blah During the Pandemic? It's called languishing," *New York Times*, December 3, 2021.

8. G. Petriglieri, "F**k science!? An invitation to humanize organization theory," *Organizational Theory* 1 (2020): 263178771989766.

9. L. Cameron, "(Relative) Freedom in Algorithms: How digital platforms repurpose workplace consent," *Academy of Management Proceedings 2021* (2021): 11061.

10. T. Chamorro-Premuzic, "Can Surveillance AI Make the Workplace Safe?," *MIT Sloan Management Review* 62 (2020): 13–15.

11. T. Chamorro-Premuzic, "This Is Why AI Will Never Be as Creepy as a Micromanaging Boss," *Fast Company*, July 21, 2020, https://www.fastcompany.com/90529991/this-is-why-ai-will-never-be-as-creepy-as-a-micromanaging-boss.

12. K. Schwab, "The Fourth Industrial Revolution: What it means and how to respond," World Economic Forum, January 2016, https://www.weforum.org/agenda/2016/01/the-fourth-industrial-revolution-what-it-means-and-how-to-respond/.

13. Petriglieri, "F**k science!?"

14. R. Vonderlin et al., "Mindfulness-Based Programs in the Workplace: A meta-analysis of randomized controlled trials," *Mindfulness* 11 (2020): 1579–1598.

15. M. Mani et al., "Review and Evaluation of Mindfulness-Based iPhone Apps," *JMIR Mhealth and Uhealth* 3, no. 3 (2015): e82, https//mhealth.jmir.org/2015/3/e82 3, e4328.

16. Lifeed, "MultiMe: How your multiple roles enrich you," July 20, 2020, https://lifeed.io/en/2020/07/20/multime-best-version-self/.

17. S. Blackburn, *Being Good: An introduction to ethics* (New York: Oxford University Press, 2001).

18. S. Pinker, *The Village Effect: How face-to-face contact can make us healthier and happier* (Toronto: Vintage Canada, 2014).

19. B. Fröding and M. Peterson, "Friendly AI," *Ethics and Information Technology* 23 (2021): 207–214.

20. V. E. Frankl, *Man's Search for Meaning* (New York: Simon and Schuster, 1985).

21. Microsoft, "Music Generation with Azure Machine Learning," Microsoft Docs, 2017, https://docs.microsoft.com/en-us/archive/blogs/machinelearning/music-generation-with-azure-machine-learning; Magenta TensorFlow, https://magenta.tensorflow.org/; Watson Beat Archives, https://www.ibm.com/blogs/research/tag/watson-beat/; Flow Machines, Sony, http://www.flow-machines.com/; AIVA—The AI composing emotional soundtrack music, https://www.aiva.ai/; Amper Music—AI Music Composition Tools for Content Creators, https://www.ampermusic.com/; L. Plaugic, "Musician Taryn Southern on Composing Her New Album Entirely with AI," *Verge*, August 27, 2017, https://www.theverge.com/2017/8/27/16197196/taryn-southern-album-artificial-intelligence-interview.

22. J. Gillick, K. Tang, and R. M. Keller, "Machine Learning of Jazz Grammars," *Computer Music Journal* 34 (2010): 56–66.

23. Books written by AI, https://booksby.ai/.

24. M. Burgess, "Google's AI Has Written Some Amazingly Mournful Poetry," *Wired*, May 18, 2016, https://www.wired.co.uk/article/google-artificial-intelligence-poetry.

25. M. A. Boden, *Artificial Intelligence: A very short introduction* (New York: Oxford University Press, 2018).

26. D. Burkus, *The Truth About How Innovative Companies and People Generate Great Ideas* (Hoboken, NJ: Jossey-Bass, 2013).

Chapter 7

1. K. E. Twomey and G. Westermann, "Curiosity-Based Learning in Infants: A neurocomputational approach," *Developmental Science* 21 (2018).

2. P. Y. Oudeyer and L. B. Smith, "How Evolution May Work through Curiosity-Driven Developmental Process," *Topics in Cognitive Science* 8 (2016): 492–502.

3. S. von Stumm, B. Hell, and T. Chamorro-Premuzic, "The Hungry Mind: Intellectual curiosity is the third pillar of academic performance," *Perspectives on Psychological Science* 6 (2011): 574–588.

4. M. Schaller and D. R. Murray, "Pathogens, Personality, and Culture: Disease prevalence predicts worldwide variability in sociosexuality, extraversion, and openness to experience," *Journal of Personality and Social Psychology* 95 (2008): 212–221.

5. C. L. Fincher and R. Thornhill, "Parasite-Stress Promotes In-Group Assortative Sociality: The cases of strong family ties and heightened religiosity," *Behavioral and Brain Sciences* 35 (2012): 61–79.

6. T. Chamorro-Premuzic and B. Taylor, "Can AI Ever Be as Curious as Humans?," hbr.org, April 5, 2017, https://hbr.org/2017/04/can-ai-ever-be-as-curious-as-humans.

7. J. Golson, "GM's Using Simulated Crashes to Build Safer Cars," *Wired*, April 2015, https://www.wired.com/2015/04/gms-using-simulated-crashes-build-safer-cars/.

8. B. Schölkopf, "Artificial Intelligence: Learning to see and act," *Nature* 518 (2015): 486–487.

9. SethBling, "MarI/O—Machine Learning for Video Games," YouTube video, June 13, 2015, https://www.youtube.com/watch?v=qv6UVOQ0F44.

10. H.-Y. Suen, K.-E. Hung, and C.-L. Lin, "TensorFlow-Based Automatic Personality Recognition Used in Asynchronous Video Interviews," *IEEE Access* 7, (2019): 61018–61023.

11. S. Nørskov et al., "Applicant Fairness Perceptions of a Robot-mediated Job Interview: A video vignette-based experimental survey," *Frontiers in Robotics and AI* 7 (2020): 163.

12. B. Russell, *A History of Western Philosophy, and Its Connection with Political and Social Circumstances from the Earliest Times to the Present Day* (New York: Simon & Schuster, 1967).

13. D. Farber, "Google Search Scratches Its Brain 500 Million Times a Day," CNET, May 13, 2013, https://www.cnet.com/tech/services-and-software/google-search-scratches-its-brain-500-million-times-a-day/.

14. T. B. Kashdan et al., "The Curiosity and Exploration Inventory-II: Development, factor structure, and psychometrics," *Journal of Research in Personality* 43 (2009): 987–998; R. Hogan, T. Chamorro-Premuzic, and R. B. Kaiser, "Employability and Career Success: Bridging the gap between theory and reality," *Industrial and Organizational Psychology* 6 (2013): 3–16.

15. M. Zajenkowski, M. Stolarski, and G. Meisenberg, "Openness, Economic Freedom and Democracy Moderate the Relationship between National Intelligence and GDP," *Personality and Individual Differences* 55 (2013): 391–398.

16. von Stumm, Hell, and Chamorro-Premuzic, "The Hungry Mind."

17. S. S. Tomkins, *Affect, Imagery, Consciousness* (New York: Springer Publishing Company, 1962).

18. Wikiquote, Albert Einstein, 2022, https://en.wikiquote.org/wiki/Albert_Einstein.

19. Bond Vililantes, "Rise of the Robots—Technology and the Threat of a Jobless Future: An interview with Martin Ford," YouTube Video, May 3, 2016, https://www.youtube.com/watch?v=Z3EPG-_Rzkg.

20. ManpowerGroup, ManpowerGroup @ World Economic Forum Annual Meeting 2022, https://wef.manpowergroup.com/.

21. J. Patrick, "Democratic Professors Outnumber Republicans 9 to 1 at Top Colleges," *Washington Examiner*, January 23, 2020, https://www.washingtonexaminer.com/opinion/democratic-professors-outnumber-republicans-9-to-1-at-top-colleges.

22. A. Agrawal, J. Gans, and A. Goldfarb, *Prediction Machines: The simple economics of artificial intelligence* (Boston: Harvard Business Review Press, 2018).

23. F. L. Schmidt and J. E. Hunter, "The Validity and Utility of Selection Methods in Personnel Psychology: Practical and theoretical implications of 85 years of research findings," *Psychological Bulletin* 124 (1998): 262–274.

24. F. Leutner, R. Akhtar, and T. Chamorro-Premuzic, *The Future of Recruitment: Using the new science of talent analytics to get your hiring right* (Bingley, UK: Emerald Group Publishing, 2022)

25. T. Chamorro-Premuzic, "Attractive People Get Unfair Advantages at Work. AI Can Help," hbr.org, October 31, 2019, https://hbr.org/2019/10/attractive-people-get-unfair-advantages-at-work-ai-can-help.

26. G. Orwell, *1984* (London: Everyman's Library, 1992).

Chapter 8

1. W. H. Bruford, *The German Tradition of Self-Cultivation 'Bildung' from Humboldt to Thomas Mann* (New York: Cambridge University Press, 2010).

2. S. Chuang and C. M. Graham, "Embracing the Sobering Reality of Technological Influences on Jobs, Employment and Human Resource Development: A systematic literature review," *European Journal of Training and Development* 42 (2018): 400–416.

3. P. Thiel with B. Masters, *Zero to One: Notes on startups, or how to build the future* (New York: Currency, 2014).

4. Quote Investigator, "History is just one damn thing after another," 2015, https://quoteinvestigator.com/2015/09/16/history/#:~:text=Some.

5. Quote Investigator, "There are only two tragedies. One is not getting what one wants, and the other is getting it," 2019, https://quoteinvestigator.com/2019/08/11/.

6. T. Chamorro-Premuzic, S. Von Stumm, and A. Furnham, *The Wiley-Blackwell Handbook of Individual Differences* (Hoboken, NJ: Wiley-Blackwell, 2015).

7. P. Rose, W. K. Campbell, and S. P. Shohov, "Greatness Feels Good: A telic model of narcissism and subjective well-being," in *Advances in Psychology Research*, vol. 31, ed. S. P. Shohov (Hauppauge, NY: Nova Science, 2004), 3–26.

8. M. J. Wilkinson, "Lies, Damn Lies, and Prescriptions," *M. Jackson Wilkinson* (blog) November 6, 2015, https://mjacksonw.com/lies-damn-lies-and-prescriptions-f86fca4d05c.

9. B. Russell, *A History of Western Philosophy, and Its Connection with Political and Social Circumstances from the Earliest Times to the Present Day* (New York: Simon & Schuster, 1967).

10. C. L. Fincher and R. Thornhill, "Parasite-Stress Promotes In-Group Assortative Sociality: The cases of strong family ties and heightened religiosity," *Behavioral and Brain Sciences* 35 (2012): 61–79.

11. L. Carroll, *Alice's Adventures in Wonderland* (Mineola, NY: Dover Books, 1993).

12. A. Smith, *The Theory of Moral Sentiments* (Erie, PA: Gutenberg Publishers, 2011).

13. C. Darwin, *The Descent of Man* (Overland Park, KS: Digireads.com, 2009).

14. Z. Schiffer, "Google Fires Second AI Ethics Researcher Following Internal Investigation," *Verge*, February 19, 2021, https://www.theverge.com/2021/2/19/22292011/google-second-ethical-ai-researcher-fired.

15. J. Faulconer, "Times and Seasons," The Quotidian, February 2006, https://www.timesandseasons.org/harchive/2006/02/the-quotidian/.

16. Wikipedia, Melvin Kranzberg, 2022, https://en.wikipedia.org/wiki/Melvin.

17. M. Ridley, "Don't Write Off the Next Big Thing Too Soon," *Times*, November 6, 2017, https://www.thetimes.co.uk/article/dont-write-off-the-next-big-thing-too-soon-rbf2q9sck.

18. "Steven Pinker: Can numbers show us that progress is inevitable?," NPR, August 17, 2018, https://www.npr.org/2018/08/17/639229357/steven-pinker-can-numbers-show-us-that-progress-is-inevitable.

19. N. Chomsky, *Requiem for the American Dream: The 10 Principles of Concentration of Wealth and Power* (New York: Seven Stories Press, 2017).

Index

addictive behavior. *See* digital addiction
ADHD-like behaviors, 26, 32, 55
advertising revenue, and consumer
    behavior data, 18–19, 20
Agrawal, Ajay, 132
AI
    algorithms in, 3
    bad behavioral tendencies unleashed
        by, 4
    capacity of, for reshaping how we
        live, 27
    claims and bleak predictions about, 1
    creativity in music creation with,
        117–118
    cultural differences eroded by, 24–25
    datafication of you and, 9, 14–18
    enablers of change by, 9
    ethical questions about, 76–81
    gathering and analyzing data on
        human behavior in, 5, 14
    human behavior changes and, 1–2, 9,
        24–27, 55
    human bias in decision-making
        versus, 3–4
    human desire to understand versus,
        135–139
    human intelligence impacted by, 9
    hyper-connected world and, 9–14
    lucrative prediction business and, 9,
        18–24
    mechanizing of intellectual work by,
        111–112
    moral character of, 76–77
    omnipresence of, in daily interac-
        tions, 3
    as opportunity for upgrading human-
        ity, 154–157

personality and, 63–65
potential for improving our lives, 3
richness of human mind and doing
    a bit of everything as limits on,
    118–119
Aibo, 78
Alphabet, 18, 23
Amazon
    business of data use and value of, 23
    desire to understand recommenda-
        tions of, 135
    hyper-connectedness during
        Covid-19 and increase in value
        of, 22
    prediction based on past data used by,
        21, 112
    recruitment AI of, 71–72
Amazon Web Services, 22
Angelou, Maya, 147
anxiety
    being unfriended on Facebook
        and, 93
    distraction and, 35
Apple
    business of data use and value
        of, 23
    hyper-connectedness during
        Covid-19 and increase in value
        of, 22
    watches sold by, and quantification of
        our data, 33
Aristotle, 47
artificial intelligence. *See* AI
artistic creation, AI in, 118
attention, 30–31
    battle for focus in, 31–32
    cocktail party effect and, 36

attention (*continued*)
  cognitive change from online media
    use and, 34
  data and, 31
  degradation of focus-related behav-
    iors in, 32
  focus on off-line activities and,
    36–37
  impact of control over, 33
  information overload and, 30–31
  messages and posts creating a state of
    hyper-alertness and, 34–35
  percentage of time online and, 32–33
  quantification of, 32, 33
attention deficit hyperactivity disorder
  ADHD–like behaviors, 26, 32, 55
authenticity (being authentic), 93–97
  curating our virtual selves and, 97
  impression management versus,
    96–97
  "just be yourself" mantra and,
    94–95
  leadership and, 95–96
authorship, AI in, 118
automation
  curiosity and new skills needed
    in, 128
  learnability as key antidote to, 128
  musical creation and, 17
  of ourselves and our lives, in
    AI age, 143
  reduced thinking in, 111–112
  unfair or unethical processes in, and
    inequality, 80

Barrett, Lisa Feldman, 62
Baumeister, Roy, 57
behavior
  AI's ability to gather and analyze data
    on, 5, 14, 50
  AI's impact on changes in, 1–2, 9,
    24–27, 55
  apps with personalized choices based
    on, 50–51
  creating new patterns of, versus pre-
    dictive algorithms, 114–116
  Facebook's manipulative tactics for
    decoding and molding, 21

  mismatch between ancient adapta-
    tions and current challenges in,
    26, 55
  social media's impact on, 151
  System 1 mode in, 53
behavioral economics
  bias and, 61–62, 63
  System 1 mode and, 53
Being Good (Blackburn), 148–149
being human, 141–157
  AI's technological revolution and,
    142–143
  automating of ourselves and our lives
    in AI age and, 143
  chasing happiness and, 144–147
  cultural differences in, 143–144
  as work in progress, 147–153
Beuys, Joseph, 118
Bezos, Jeff, 7, 91
bias, 61–82
  AI algorithms matching content
    to, 62
  AI's exposing of, 69–70
  approaches to becoming less biased,
    67–68
  awareness gap about our own biases,
    66–67
  decision-making and, 3–4, 68–73
  human endowment with range of,
    61–62
  irrationality patterns and, 63
  optimism in predictions and, 66
  problems with workplace training on,
    66–67
  rational and objective behavior
    versus, 63
  social media and, 62
  testing yourself for your level of, 82
Big Tech. *See* tech firms
Blackburn, Simon, 148–149
Boden, Margaret, 119
body language, in virtual settings, 76
Bohr, Niels, 115
booksby.ai (online bookstore), 118
Brexit referendum, AI tools in, 80
"broadcast intoxication"
  phenomenon, 89
Buffet, Warren, 52

Bukowski, Charles, 119
Bumble, 50

Cambridge Analytica, 79, 80, 128
cancel culture, 129
Carr, Nicholas, 34
Carroll, Lewis, 149
children
    screen time limits and, 41–42
    inborn curiosity of, 121
China, behavioral surveillance in, 20
choice paradox, 105
Chomsky, Noam, 156–157
Christie's, 118
Churchill, Winston, 91
cocktail party effect, 36
cognitive change, and online media
    use, 34
Coltrane, John, 37
competitiveness need, 11, 12
confirmation biases, and social
    media, 62
consumption patterns
    choice paradox and, 105
    obsession with happiness and, 144
    online shopping and, 51–52
Costa, Petra, 81
Covid-19 pandemic
    curiosity and risks from socializing
        during, 122
    digital friendships during, 116–117
    digital hyper-connectedness during,
        12–13, 21, 41, 143
    increase in online shopping
        during, 22
    vaccine inequalities and, 150
creativity
    AI in music creation and, 117–118
    being creative online, versus predic-
        tive algorithms, 114–116
    ways of boosting, 119
cultural differences
    AI's erosion of, 24–25
    being human and, 143–144
curiosity, 121–140
    AI algorithms for asking relevant
        questions in, 126–127
    AI substitutes for, 122–125

desire to understand and, 135–139
employability and, 127–128
game-playing and, 123–124
human intelligence in AI age and,
    131–132
inborn human instinct for, 121
openness and, 128–129
survival instincts and, 122
testing yourself for your level of, 140
ways of reclaiming, 127–131

Darwin, Charles, 151
Darwinism, 151
data
    attention and, 31
    as real product of being in a hyper-
        connected world, 13–14
    use of as world's biggest business, 23
datafication, 9, 14–18
    actions generating a repository of
        digital signals in, 15
    increase in online shopping during
        Covid-19 and, 22
    perceived value of tech firms related
        to, 17–18
    predictions about behavior and ac-
        tions based on, 15–16, 136
dating sites
    AI's improving of results on, 69, 70
    desire to understand matches of,
        135–136
    in-person encounters replaced by,
        50, 51
    prediction and, 20
Davis, Miles, 37, 118
decision-making
    AI's insights in recommendations
        and, 136
    AI's reducing our need to think and
        make choices in, 105–106, 109
    ethical questions about machines
        replicating, 77–78
    human bias in, versus AI, 3–4, 68–73
    impact of AI-enabled distractions
        on, 35
Declaration of Independence, 146
depression, and being unfriended on
    Facebook, 93

Dhawan, Erica, 76
digital addiction
    application of AI to increase, 53
    impatience and impulsivity levels and
        propensity for, 48–49
    new normal of digital compulsions
        and, 49–50
    psychiatric pathology of, 49
Digital Body Language (Dhawan), 76
distraction, 29–46
    approaches to limiting, 37–38
    attention economy and, 30–31
    decision-making and, 35
    estimated economic cost of, 39
    examples of, in daily life, 29–30
    finding meaning and, 43–45
    focus on off-line activities and,
        36–37
    health and behavior impact of, 35–36
    increasing levels of screen time
        and, 42
    levels of anxiety and stress related
        to, 35
    messages and posts creating a state of
        hyper-alertness and, 34–35
    percentage of time online and, 32–33
    percentage of time working and,
        39–40
    recovery time from, 38–39
    sensory overstimulation and, 44
    testing yourself for your level of, 46
Durant, Ariel, 8, 99
Durant, Will, 8, 25, 99

echo chamber effect, 54, 75
e-commerce. *See* online shopping
economic innovation, and social evolu-
    tion, 8
Economist, 23, 38, 39
Edmonson, Amy, 100
Einstein, Albert, 128
elections, AI tools in, 80
Emin, Tracey, 118
ethical questions, 76–81
    Facebook's identification of undecid-
        ed voters and, 79–80
    genetic profiling and, 78–79
    legal matters and, 81
    machines replicating human
        decision-making and, 77–78
    moral character of AI and, 76–77
    political targeting and, 80
evolution, and innovation change, 8
exercise, and self-control capacity, 58

Facebook
    academic performance related to
        amount of time on, 34
    ad revenue and consumer behavior
        data of, 18–19
    approaches to limiting distractions
        from, 37–38
    being creative on, versus predictive
        algorithms, 114, 116
    business bet on data and WhatsApp
        acquisition by, 16–17
    constraints on the range of responses
        or behaviors displayed in, 15
    curating our virtual selves on, 97
    depression and anxiety increases
        from being unfriended on, 93
    expressing and fulfilling our universal
        needs using, 12
    hyper-connectedness during Covid-19
        and increase in value of, 22, 23
    identification of undecided voters by,
        79–80
    increase in use of during Covid-19,
        41
    impatience or impulsivity levels and
        propensity to be addicted to, 48
    likes and dislikes on, 74
    manipulative tactics behind algo-
        rithms of, 21
    narcissistic tendencies in use of, 87, 88
    predictions about personality and
        values based on choices in, 15–16
fear of missing out. *See* FOMO
feedback
    likes and dislikes for, 74
    managers and, 74–75
Fiverr, 50
FOMO (fear of missing out)
    increasing levels of screen time and,
        42, 43
    online shopping and 51

Ford, Henry, 105
Frankl, Viktor, 117
free will, 109–110

Galloway, Scott, 105
game-playing, curiosity in, 123–124
Gandhi, 147
Garrix, Martin, 93
Gates, Bill, 1, 90
genetic profiling, 78–79
gig workers, 50, 110–111
God complex, 90–91
Goffman, Erving, 94
Google, 152
    ad revenue and consumer behavior
        data of, 18
    AI algorithms for asking relevant
        questions in, 126
    being creative in searches on, versus
        predictive algorithms, 114
    human curiosity and, 123
Google Maps, 105
Grant, Adam, 109

happiness, 144–147
    narcissistic aspect of, 145
    personality traits as predictor of, 145
    positive psychology on, 144
Harari, Yuval, 13
Hari, Johann, 33
Hawking, Stephen, 1, 74
Heidegger, Martin, 119
HireVue, 50
hiring process
    AI evaluation of résumés and inter-
        views in, 3, 124–125
    AI platforms and, 50, 51, 69, 70–72
    candidate curiosity and learnability as
        factors in, 127–128
    recruiters' use of candidate data from
        LinkedIn in, 20
Hobbes, Thomas, 24
humanity, 141–157
    AI's experience of, 141
    AI's revolution as opportunity for
        upgrading, 154–157
    automating of ourselves and our lives
        in AI age and, 143

cultural differences in, 143–144
effects of AI's technological revolu-
    tion on, 142–143
as work in progress, 147–153
Humboldt, Alexander von, 141
humility, 98–101
    advantages to, 98–99
    human intelligence in AI age and,
        131–132
    leaders and, 99–100
hyper-connected world, 9–14
    competitiveness need and, 11, 12
    Covid-19 and, 12–13, 21, 143
    current behaviors and preexisting
        human desires in, 10
    data as real product of, 13–14
    as defining feature of AI age, 9–10
    desire to connect with each other
        in, 11
need to find meaning and,
    11–12
relatedness need and, 11, 12

I Am AI (Southern), 117
impulsivity
    impact of, 54
    lack of patience and, 48
    mass media and, 55
    propensity to be addicted to AI plat-
        forms and levels of, 48–49
    social media reactions and, 68
industrial revolution, 1, 111, 142
Inge, William, 142
innovation
    curiosity and, 123
    discontent and unhappiness as moti-
        vators for, 146–147
    God complex and, 90–91
    history of blaming for cultural demise
        and degeneration, 5
    human interest in going faster and, 48
    psychological consequences of, 8
    technological unemployment from,
        142
In Search of Lost Time (Proust), 30
Instagram
    ad revenue and consumer behavior
        data of, 18–19

Instagram (*continued*)
    competition for our attention and
        focus by, 32
    datafication of you and, 17
    impatience or impulsivity levels and
        propensity to be addicted to, 48
    increase in use of during Covid-19, 41
    likes on, 74
    sensory overstimulation and, 44
intellectual innovation, and social
    evolution, 8
intelligence, AI impact on, 9
internet of things, 13
Ishiguro, Kazuo, 13–14

job creation, in automation, 128,
    142–143
job interviews
    AI evaluation of, 3, 124–125
    being the best version of you
        in, 94
    structured questions in, 133–134
job searches, and AI platforms, 50, 51,
    136
job skills, curiosity, and learnability as,
    127–128

Kant, Immanuel, 81
Kardashian, Kim, 92
Kranzberg, Melvin, 154

leaders
    being authentic and, 95–96
    humility and, 99–100
Leadership BS (Pfeffer), 95
learnability, as a job skill, 128
learning, curiosity in, 127–128
Lessons of History (Durant and Durant), 8
LinkedIn
    being creative on, versus predictive
        algorithms, 116
    curating our virtual selves on, 97
    endorsements on, 74
    expressing and fulfilling our universal
        needs using, 12
    job searches using, 50
    recruiters' use of candidate data from,
        20

managers
    control over workers and, 110–111
    feedback provided by, 74–75
ManpowerGroup, World Economic
    Forum, 128
Match Group, 20
meaning
    AI age and relationship with, 44
    need to find, 11–12, 43
    removing ourselves from digital sur-
        plus of information and, 44–45
Meta, 17, 18–19
Meyer, David, 36
Microsoft, 71
    business of data use and value of, 23
    hyper-connectedness during Covid-19
        and increase in value of, 22
mindfulness apps, 112
Molière, 47
Montagu, Ashley, 138
moral innovation, and social
    evolution, 8
multitasking, and productivity, 39
musical preferences, human under-
    standing of versus Spotify, 136–137
music creation
    AI in, 117–118
    datafication and, 17
Musk, Elon, 1, 43, 85, 90, 91

narcissism, 83–102
    AI age's normalization of, 84
    "broadcast intoxication" and, 89
    dependence on others to inflate our
        egos in, 93
    digital technologies' relationship to,
        84
    exceptional achievers and, 90–91
    generational changes in measures of,
        84–85
    grandiose exhibitionism in, 85
    happiness and, 145
    humility as an antidote to, 98–101
    metrics of engagement and, 89
    psychiatric diagnosis of, 84
    selfies and, 87, 88
    social media use and, 85–88
    testing yourself for your level of, 102

Narcissus myth, 83
Netflix
    algorithms' conclusions about activity
        on, 138
    algorithms for recommendations on,
        51
    being creative in choices on, versus
        predictive algorithms, 114, 116
    competition for our attention and
        focus by, 32
    datafication of you and, 17
    desire to understand recommenda-
        tions of, 135
    increase in use of during Covid-19,
        41
    prediction based on past data used by,
        21, 105, 112
Neuralink, 43
Newton, Isaac, 47
Nielsen survey, 42
1984 (Orwell), 137

online shopping, 51–52
    AI sites with sticky relationships in,
        51
    compulsive consumption patterns in,
        51–52
    increase in during Covid-19, 22
openness, 128–129
optimism bias, 66
Orwell, George, 137
Oswalt, Patton, 62
Otaku culture, Japan, 25

patience, 47–60
    advantages to boosting, 56–57
    AI's impact on, 49–50
    consuming actual information with,
        54
    historical belief in power of, 47
    impatience and propensity to be
        addicted to AI platforms, 48–49
    impulsivity and lack of, 48
    need to react quickly and, 53–54
    testing yourself for your level
        of, 60
    wide range of choices on AI apps and
        need for, 51

personality
    bias and irrationality patterns in, 63
    possible AI acquisition of, 63–65
    as predictor of happiness, 145
Peterson, Jordan, 93
Petriglieri, Gianpiero, 112
Pfeffer, Jeffrey, 95
political innovation, and social evolu-
    tion, 8
political targeting, AI tools in, 79–80
Portrait of Edmond de Belamy
    (AI-generated work), 118
positive psychology, 144
predictable machines, 103–120
    AI age's turning humans into,
        103–104
    AI's reducing our need to think and
        make choices as, 104–107
    being creative in online choices, ver-
        sus predictive algorithms, 114
    being more creative in your personal
        life versus, 119
    creating diverse versions of ourselves
        and new behavior patterns versus,
        114–116
    diluted breadth and depth of experi-
        ence in, 107
    free will and, 109–110
    historical assumption about human
        complexity versus, 107–108
    mechanizing of intellectual work by
        AI and, 111–112
    richness of human mind and doing a
        bit of everything versus, 118–119
    serendipity and, 113–119
    testing yourself for acting like, 120
    understanding who is in control and,
        109–113
prediction, 18–24
    algorithms' conclusions about behav-
        ior in, 138
    AI identification of patterns and, 132
    AI monetization of patterns of our
        habits and behaviors in, 19–20
    being creative in online choices, ver-
        sus algorithms for, 114, 116
    business of data use and value of tech
        companies in, 23

prediction (*continued*)
  China's citizen management system
    using, 20
  control related to, 109
  datafication of you and, 15–16
  dating sites using, 20
  desire to understand and, 135–139
  Facebook's identification of undecid-
    ed voters and, 79–80
  humans as predictable machines and,
    107
  lucrative business of, 9, 18–24
  optimism bias in, 66
  recruiters' use of LinkedIn's data for,
    20
  selling targeted content and personal-
    ized ads using, 18–19, 20
  as surveillance capitalism, 20–21
Prediction Machines (Agrawal, Gans,
    and Goldfarb), 132
presidential elections, AI tools
    in, 80
productivity
  approaches to avoiding distractions
    for improving, 37
  multitasking and, 39
  percentage of working time during
    distractions and, 39–40
  recovery time from digital interrup-
    tions and, 38–39
  working from home and, 40
Proust, Marcel, 30
psychological consequences of techno-
    logical innovations, 8
PWC, 18

Reagan, Ronald, 81
recruitment
  AI evaluation of résumés and inter-
    views in, 3, 124–125
  candidate data from LinkedIn used
    in, 20
  gender differences in AI candidates
    for, 71–72
relatedness need, 11, 12
Robson, Gregory, 44
Rose, Frank, 25
Rousseau, Jean-Jacques, 24

Schopenhauer, Arthur, 65
Schubert, Franz, 118
scientific management, 110–111
search engines
  AI algorithms for asking relevant
    questions in, 126–127
  AI's improving of results on, 69
self-control
  capacity to develop higher levels of,
    57
  exercise and, 58
  sleep quality and quantity and, 58
self-driving autonomous cars, 106, 111
self-esteem, and social media feedback,
    89–90, 146
selfies, 87, 88
serendipity, versus predictive algo-
    rithms, 113–119
Shallows, The (Carr), 34
Sharot, Tali, 66
shopping. *See* online shopping
Simon, Herbert, 30–31
sleep, and self-control capacity, 58
smartphones
  addiction to, 49
  productivity and distractions from,
    38, 39, 40
Smith, Adam, 149, 151
Snapchat, 33, 74, 87
Social Dilemma (documentary), 21
social media
  academic performance related to
    amount of time on, 34
  behavioral impact of, 151
  being creative in choices on, versus
    predictive algorithms, 114, 116, 117
  "broadcast intoxication" on, 89
  competitiveness need and, 12
  confirmation biases and, 62
  curating our virtual selves on, 97
  as a distraction machine, 32
  health and behavior impact of screen
    use in, 35–36
  impulsive reactions to people's actions
    on, 68
  messages and posts creating a state of
    hyper-alertness in, 34–35
  narcissism and, 85–88

need to find meaning and, 12
   relatedness need and, 11, 12
   school distractions due to use of, 34
self-esteem and feedback on, 89–90,
   146
   selfies on, 87, 88
Socrates, 5
Sony, 78
Southern, Taryn, 117
Spotify, 17, 29, 41
   desire to understand recommenda-
     tions of, 135
   human understanding of musical
     preferences versus choices of,
     136–137
   prediction based on past data used
     by, 21
Stockholm syndrome, 107
Summers, Larry, 16
Super Mario World (game), 124
surveillance capitalism, prediction business as, 20–21
System 1 mode, 53

Target, 19
Taylor, Frederick, 110–111
tech firms
   ad revenue and consumer behavior
     data of, 18–19
   business of data use and value of, 23
   hyper-connectedness during Covid-19
     and increase in value of, 22
   perceived value of based on data
     collected by, 17–18
technological innovations. *See* innovation
Thiel, Peter, 10
TikTok
   expressing and fulfilling our universal
     needs using, 12
   likes on, 74
   narcissistic tendencies and use of, 87
   personalized recommendations to
     hook consumers on, 52–53
Tinder, 20, 50, 104
Tolstoy, Leo, 47
Trump, Donald, 80, 150
Twenge, Jean, 84

23andMe, 78, 80
Twitter, 71
   algorithms' conclusions about activity
     on, 138
   being creative on, versus predictive
     algorithms, 114
   competition for our attention and
     focus by, 32
   curating our virtual selves on, 97
   echo chamber effect on, 54
   likes on, 74
   Musk's acquisition deal for, 85
   retweet predictions based on data
     collected by, 16
   speedy interactions with others
     on, 54

Uber, 17, 50, 109
understanding, desire for, 135–139
Upwork, 50

valuation of firms, based on data
   collected, 17–18
Visser, Margaret, 154
Vonnegut, Kurt, 86

Wall Street Journal, 53
WhatsApp
   ad revenue and consumer behavior
     data of, 18–19
   datafication of you and, 17
   Facebook's business bet on data and
     acquisition of, 16–17
   impatience or impulsivity levels and
     propensity to be addicted to, 48
*Why Do So Many Incompetent Men
   Become Leaders?* (Chamorro-
   Premuzic), 99
Wilde, Oscar, 145
working from home, and productivity,
   40
workplace
   AI revolution's impact on tasks in,
     142–143
   bias training in, 66–67
   curiosity and learnability in,
     127–128
   dehumanization of, 112

workplace (*continued*)
  mechanizing of intellectual work by
    AI in, 111–112
  tracking software in, 32
  understanding who is in control and,
    110–111
World Economic Forum, 128

"You of Things" concept, 13
YouTube

being creative in choices on, versus
  predictive algorithms, 116
increase in use of during Covid-19,
  41

Zezza, Riccarda, 114
Zoom, increase in use of during
  Covid-19, 13, 41
Zuboff, Shoshana, 20–21
Zuckerberg, Mark, 74

Acknowledgments

This book is the product of many discussions, exchanges, and conversations with brilliant minds, who, over the years, have had a profound impact on my views on the human-AI interface. I am grateful to my *Harvard Business Review* editors, Dana Rousmaniere, Paige Cohen, and Sarah Green Carmichael (now at Bloomberg), for helping me shape my early ideas on this subject, as well as Kevin Evers, the editor of this book, for enduring the near-masochistic process of curating, editing, and especially cleaning and cleansing the ramblings and ruminations he received from me, and magically turning them into this book. Thanks also to Lydia Dishman at *Fast Company* for always asking interesting questions, which help push my thinking.

I am also grateful to the people who always inspire—and help—me to bridge theory and practice in the human-AI field. My amazing ManpowerGroup colleagues, especially Becky Frankiewicz, Michelle Nettles, Ganesh Ramakrishnan, Stefano Scabbio, Francois Lancon, Alain Roumihac, Riccardo Barberis, and Jonas Prising, for their commitment to harness ethical AI to help millions of people thrive at Leutner, Reece Akhtar, Uri Ort, Gorkan Ahmetoglu, for turning many of these ideas into innovations, and bridging the divide between AI and IO (psychology). My amazing coauthors, thought partners, and idea instigators, especially Amy Edmondson, Herminia Ibarra, Cindy Gallop, Katarina Berg, Nathalie Nahai,

Darko Lovric, Gianpiero Petriglieri, Josh Bersin, Yuval Harari, Scott Galloway, Oliver Burkeman, and Melvyn Bragg.

Finally, I would like to thank my literary agent, Giles Anderson, for his sage advice and guidance throughout the creation of this book, which, as with my previous three books, has significantly improved what you have in front of you now.

About the Author

Tomas Chamorro-Premuzic is an international authority in psychological profiling, talent management, leadership development, and people analytics. His commercial work focuses on the creation of science-based tools that improve organizations' ability to predict performance and people's ability to understand themselves. He is currently the chief innovation officer at ManpowerGroup, a cofounder of Deeper Signals and Metaprofiling, and a professor of business psychology at both University College London and Columbia University. He has previously held academic positions at New York University and the London School of Economics and lectured at Harvard Business School, Stanford Business School, London Business School, Johns Hopkins, IMD, and INSEAD, as well as being the CEO of Hogan Assessment Systems.

Dr. Tomas has published 11 books and over 200 scientific articles on the psychology of talent, leadership, innovation, and AI, making him one of the most prolific social scientists of his generation. His work has received awards by the American Psychological Association, the International Society for the Study of Individual Differences, and the Society for Industrial-Organizational Psychology, of which he is a fellow. Dr. Tomas is also the founding director of University College London's Industrial-Organizational and Business Psychology program, and the Chief Psychometric Adviser to Harvard's Entrepreneurial Finance Lab.

Over the past twenty years, he has consulted to a range of clients in finance (e.g., JP Morgan, HSBC, Goldman Sachs), advertising (e.g., Google, WPP, BBH), media (e.g., BBC, Red Bull, Twitter, Spotify), retail (e.g., Unilever, Walmart, Tesco), fashion (e.g., LVMH, Net-a-Porter, Valentino), governmental (e.g., British Army, Royal Mail, NHS), and intergovernment organizations (e.g., United Nations and World Bank).

Dr. Tomas's media career comprises over a hundred TV appearances, including on the BBC, CNN, TED, and Sky, and over four hundred articles in *Harvard Business Review*, the *Guardian*, *Fast Company*, and *Forbes*, among others. Dr. Tomas is also a keynote speaker for the Institute of Economic Affairs. He was born and raised in the Villa Freud district of Buenos Aires but has spent most of his professional career in London and New York. His previous book is *Why Do So Many Incompetent Men Become Leaders?* (Harvard Business Review Press, 2019).